THE SEVEN CHURCHES

Revelation to The Church Today

Rev 3:22 He who has an ear, let him hear what the Spirit says to the churches.

By Ralph Brandt

Contents

THE SEVEN CHURCHES 1

DEDICATION...3

INTRODUCTION 5

THE SEVEN CHURCHES 5

EPHESUS 11

SMYRNA 27

PERGAMOS...35

THYATIRA 46

SARDIS 68

PHILADELPHIA 82

LAODICEA...93

SILENCING THE PROPHETS 106

DIVISION IN THE CHURCH..124

ABOUT THE AUTHOR 128

DEDICATION

This book is dedicated to the memory of those who have over the years spoken God's word into my life and to some who are still ministering. Some like Theodore, Cathy and George were there each for nearly 10 years, some like Eugene had only a few months and I can look back and see things he impacted for good. All had profound influence. Many of those have been promoted and are in a home in glory.

My life and faith have been shaped by many things. One is the excerpts from the song.

They tore the old country church down[1] Built a big new church way up town.

...

There is nothing wrong with the church way up town It's the members that left God down If they would humble themselves and pray They would hear from God today In that big new church way up town.

---

[1] Song by Buddie Starcher

I have worshipped in everything from a tiny home, to a MASH Tent, to a country church, to a big church, to a tent that filled a race track infield, to a cathedral. God is still God. It is the

church that makes the difference, not the building, the church which the song correctly calls the members (of the Body of Christ).

It is we who make it happen, not the building. I am including on these pages several pictures of three places of worship, Paradise Lutheran Church at Holtschwam, York County PA – an active church, the Mt. Tabor Church AME just outside Mt. Holly Springs, Cumberland County PA and Oral Roberts tent from the 1950's that seated about 16,000. I was never inside either church, was in the tent but all have had an impact on my life.

Mt. Tabor was built in about 1867 by freed slaves that came north from the Roanoke area and settled there. Sometime in the 1970's services ended. These pictures were taken in 2009. I had the privilege of hearing their last pastor preach in another church. I had the privilege to hear Oral Roberts too.

INTRODUCTION

This book started as chapters in another book that started as a study of the book of Titus and then I felt it good to separate them into two books because the study of Titus needs to stand alone as does this one on the seven churches.

The following is a paragraph from 'Titus – A Man With a Mission' which I feel is good to include here.

This book was written with two goals. The first was to show a way of studying the scriptures by following the leading of the Holy Spirit. I am not the teacher. He is. It is put down as I studied, with rabbit trails as I was led to look at another scripture, a song, an experience, or what was brought to my mind by the Holy Spirit. This is the way I learned to have the scripture; the written word become real to me. If as you read your mind is taken to some other scripture, stop reading this book and take that rabbit trail. I trust the Holy Spirit to guide you and to teach you more than you could possibly learn from me.

THE SEVEN CHURCHES

This study started with Titus and as I was working on it, I found places where pieces of the letters to the seven churches that the Angel gave to John had a significant message to the current church. But then all scripture is profitable for us. The more I read, the more I started seeing places where we today are missing the wisdom in those letters to the churches. Most preachers seem to have two hot button scriptures they pull from them and ignore the remainder of the two chapters. I will be sharing what I believe God has given me through these letters. I will set the stage with the introduction to the letters that appears in the last two verses in Revelation Chapter 1. It is the angel's instruction to John and tells clearly what is to come.

Rev 1:19 Write the things which you have seen, and the things which are, and the things which shall be after this,

In another scripture that writer is told.

Hab 2:2 And the LORD answered me, and said, Write the vision, and make it plain upon tables, that he may run that readeth it.

Hab 2:3 For the vision is yet for an appointed time, but at the end it shall speak, and not lie: though it tarry, wait for it; because it will surely come, it will not tarry.

Why write? Writing has some permanence. If the message is passed verbally it can change with each telling. With some caveats the written word is permanent. Two issues include the copying of it which was not perfect however as we have found older texts, the Dead Sea Scrolls for example, we have learned there are a few copying errors but no significant changes. The second issue is language translation. Anyone with even a superficial knowledge of more than one language knows translating is an inexact process. It is very difficult to translate and get the full meaning. Here we English speakers have an advantage, we have a wealth of translations done over about 500 years. I can choose from the KJV, MKJV, Beck Williams, NIV, NAS, Good News, Amplified, and others and with these resources, read a verse as it was translated by different translators. I have done this and because of the consistency of the translations, I have confidence in all the translations I mentioned here. Yes, I know there are some that hold to one or another translation as being the only one, but I believe they are missing an opportunity to see the scriptures be more open to them. The scripture says that out of a multitude of counsel there is safety. I believe the same holds for multiple translations.

At the time of John's vision, writing was the way of conveying ideas. Once written one could hand the document to someone to carry. But more important for scripture, it was preserved as it was revealed. Those preserved writings are the basis of our current knowledge of God, the history of Israel, the story of the early church and critically, the story of Jesus.

Over the years this story has been copied, translated into many languages, and since Gutenberg; printed. The most prolific writers of the New Testament were Paul, John and Luke. Paul's writings were to the churches and individuals and brings a significant knowledge of God and the New Covenant. Each of these and the other writers bring something to us.

In the Old Testament I have done a lot of reading in the minor prophets, particularly Amos and Habakkuk. They were the 'unofficial' or maybe better, 'unrecognized' prophets. It is probably because of the roles God has placed me in are somewhat like those they filled. An interesting scripture from that book is in the second chapter.

> Hab 2:4 Behold, his soul which is lifted up is not upright in him: but the just shall live by his faith

Hundreds of years before Christ he wrote, "the just shall live by his faith." I believe Habakkuk could look into the future and see Christ, see the new covenant and see redemption. Maybe it was not a clear picture, but it told the story. Paul says, "now we see through a glas darkly."

I also identify with Amos.

> Amo 7:13 But prophesy not again any more at Bethel: for it is the king's chapel, and it is the king's court.

> Amo 7:14 Then answered Amos, and said to Amaziah, I was no prophet, neither was I a prophet's son; but I was an herdman, and a gatherer of sycomore fruit:

> Amo 7:15 And the LORD took me as I followed the flock, and the LORD said unto me, Go, prophesy unto my people Israel.

> Amo 7:16 Now therefore hear thou the word of the LORD: Thou sayest, Prophesy not against Israel, and drop not thy word against the house of Isaac.

> Amo 7:17 Therefore thus saith the LORD; Thy wife shall be an harlot in the city, and thy sons and thy daughters shall fall by the sword, and thy land shall be divided by line; and thou shalt die in a polluted land: and Israel shall surely go into captivity forth of his land.

Amos has the word of the Lord to present to the people. He is told in verse 13 to not prophecy – to speak the word of the Lord in this location. He is being told, "go somewhere else to give God's word."

God's word can be encouragement, correction, doctrine, and warning. The church has gotten this idea that you must fit a certain mold to be presenting the word of the Lord. By that we have been guilty of silencing the prophet, the prophetic word of God. I have included later in this book a chapter on that subject.

Rev 1:20 the mystery of the seven stars which you saw in My right hand and the seven golden lampstands. The seven stars are the angels of the seven churches, and the seven lampstands which you saw are the seven churches.

The angel here explains the seven stars and the seven golden lampstands that are mentioned earlier in Chapter 1. He has set the stage. There is much controversy over whether these are existing physical churches in John's time, figurative churches, stages of the church or churches to come. I have even seen this generate frivolous contention. I have seen too many spend the time on this frivolous pursuit and ignore the wisdom available and directed to the current church in the world as its factions go off into various elements of error.

Although I believe the writings were to seven churches of that time, too much of what is discussed in these letters could be of value today if it were properly applied rather than waste them in idle speculation. It is the difference between Theology that deals in the abstract and Applied Theology which is practical and useful - what God desires we pursue. It is how we should honor God with our daily lives. The Pharisees engaged in Theology, they took it to distraction, and Jesus was not complementary of them.

The church has for too long majored on the minors, on things like the color of the church door – in one denomination they must be one color, the type and style of music that should accompany worship or even if there should be music, the order of the service, whether to use responsive readings, whether it should be trespasses or debts in the Lord's Prayer, the exact wording of any reading or song, and on and on to distraction and worse contention and strife.[2] I am becoming convinced that some people sit up all night trying to find something to use to create contention. There is a faction that has a problem with "lead us not into temptation" and wants to use, "do not let us be led into temptation". Although I have always used the former, I actually like the latter because I believe it defines God

---

[2] I cover these I detail in my book, "Destructive Christian Doctrines."

better, but I am not ready to create another denomination or a dozen over it. There are some churches that will not use the Apostle's Creed for any of twenty lame reasons in spite of the fact that it is a good reminder of what God has done for us. It is a great tool to keep the important portions of the faith. Most of the flack is over the 'holy catholic church', which they do not understand is Christ's world-wide, universal church, not the Roman Catholic Church. Some churches avoid this by using the word universal rather than catholic. I like that because it makes the message plain, avoids the confusion with the apostate Roman Catholic Church and avoids the contention. But there are others who refuse it because it is 'liturgy'. Some refuse to consider anything that came from the historical churches as valid.

I ask a couple questions here. Is "Rock of Ages", "A Mighty Fortress" and the like valid? They too come from the historic churches. Because these churches have at times lost their way and strayed from the focus on God, is that a valid reason to reject all they ever created? Like Paul I say, "God forbid." As Bill Wilson from Teen Challenge said some years ago, "chew the meat and spit the bone." I have seen two pastors effectively use the Apostle's Creed to teach basic Christianity by using it, phrase by phrase. If I were teaching 10 to 21 year old students I would press them to memorize it. I might even do that with octogenarians. For the detractors, tell me what is wrong with it. There is in my opinion one phrase to be added to it, "the inerrancy of the scriptures." I will not throw out what is very good because it is not perfect.

Let's cut to the chase. The only things that the church should be concerned with are bringing people into a saving knowledge of Christ and making disciples, those disciplined and trained to spread his word and also caring for those in need. Sure, there must be structure for doing that, the disciples recognized that when they appointed Stephen and the others to care for needs. Some of those churches that decry those churches that worship with form and ceremony have just as ingrained forms and ceremonies of their own creation!

I grew up in a Pentecostal Church that decried the formality. They preached that the "formal churches" should write Ichabod across the door lintels and cited this scripture.

> 1Sa 4:21 And she named the child Ichabod, saying, The glory is departed from Israel: because the ark of God was taken, and because of her father in law and her husband.

But these Pentecostal Churches were formal, they had a form of worship. It wasn't written, it wasn't in a book of discipline, but it was there. I could go to church, confident in knowing that we would open with a prayer, have three hymns, requests for prayer, a time of prayer, a time for testimonies – which were all too often times of complaint on how tough some had it rather than a report on God's goodness, possibly a couple special numbers, a sermon, an alter call and a benediction. We decried the bulletin and order of the service, but we were formal too. And I could go to ten of these "independent" churches and see almost exactly the same form!

I am going to throw out something and let you decide. This back woods Pentecostal boy who was raised in the fire of God, can't stand the smoke, loves God, speaks in tongues, yes, does, now is in one of those "formal churches". God put me there. Maybe God has me there to temper me, and I am not sure he didn't do that as an act of humility. He has taught me that there is only one way to God, Jesus, but there are many manifestations of worship. I can get excited listening to good worship music.

I recently bumped into Ernie Haas and Signature Sound and the Gaither Vocal Band singing Holy Highway (YouTube) and played it at least a dozen times over a couple days. As an aside, this is a performance, but I can worship the King with it. It may be a performance to some, it is worship to me if my attitude is right.

I love "When we all get to Heaven", "Onward Christian Soldiers" and a bunch of others that are rousing calls to move and I need that at times. But I can easily find God in the quiet "I Love Him", "It is well with my Soul", and "God on the Mountain" and sometimes I need that quiet time to move myself into his presence. A few years ago, New Creation Community Church moved to a Sunday AM two service format. The eight AM is a Contemplative service, the ten fifteen, Contemporary. I started in the later service, thinking I would fit better there. Before the one service was a blended service with both styles. A couple weeks I had a Search and Rescue team training and needed to be out before the end of the second service, so I attended the early, "Traditional" service. I found that time to be quiet with God was working for me, now I am there unless I either stay for both services or my schedule dictated coming to the later one.

I took a break from writing this to proof read a book by William Brandt on Worship. It will likely be published in the same time frame as this one. Ironically one who lives and breathes contemporary worship has many of the same concerns I do. I did not find a single point of significant disagreement.

We need to be careful to not put ourselves into the "They have killed all the prophets and I am the only one left" mode.

EPHESUS

Rev 2:1 To the angel of the church of Ephesus write: He who holds the seven stars in His right hand, who walks in the midst of the seven golden lampstands, says these things.

If we are to evaluate a doctrine or a passage, we need to know who is stating it. Who is this who holds the seven stars and walks in the midst of the seven lampstands? We can see that if we look back a few verses.

> Rev 1:10 I came to be in the Spirit in the Lord's day and heard behind me a great voice, as of a trumpet,

> Rev 1:11 saying, I am the Alpha and Omega, the First and the Last. Also, What you see, write in a book and send it to the seven churches which are in Asia: to Ephesus, and to Smyrna, and to Pergamos, and to Thyatira, and to Sardis, and to Philadelphia, and to Laodicea.

In most cases the Bible has the explanation, the answer, if we are wise enough to look for it and astute enough to recognize it. He clearly gives his credentials. "I am the Alpha and Omega, the First and the Last." This one is none other than the one who was there in the beginning. He is the second portion of the Godhead, Jesus Christ who was there. The scripture states clearly that Jesus was there in the beginning. We have the Father, the Son and the Holy spirit, the Triune Godhead, the three in one, the Trinity.

> Joh 1:1 In the beginning was the Word, and the Word was with God, and the Word was God.

11

He is the Word that was made flesh and dwelt among us.

> Joh 1:14 And the Word was made flesh, and dwelt among us, (and we beheld his glory, the glory as of the only begotten of the Father,) full of grace and truth.

> Joh 1:15 John bare witness of him, and cried, saying, This was he of whom I spake, He that cometh after me is preferred before me: for he was before me.

John the Baptist attested to this and those who today know the power of His forgiving love and His blood know. All three of the Trinity were there.

I know, for some I committed heresy, the word Trinity is not in the King James Version of the Bible, so this word is an anathema to those who call it the only real version of the bible. Let me address that. The trinity, the Father, Son and Holy Spirit is in the KJV, it just isn't called that because the translators did not use that word. Does the name matter when it is a valid description of the Triune God Head? The big question is, how can there be three in one? We fail to understand that because we are stuck in our mental trap and not allowing God's word to speak to us. We major on the minors and play games at the foot of the cross while people are literally going to hell. We argue over words rather than doing what things need to be done. We are worse than the Roman Soldiers who cast lots for his garments. They did not know what they were doing!

Any doubt of this should be cleared up with the first verses in the gospel of John. Christ was the word that was there when this all began. I have pulled a couple verses, but I challenge you to read that chapter at least a couple times. Read it and savor it like you would a good cup of tea.

We should also note that he states clearly this is to the seven churches in Asia. Although that is true, I find what is written to them applies to many churches today. I see things in more than one of the churches that apply to individual churches and denominations today. We fail to see the value of these chapters if we treat this only as history just as much as we do if we treat the whole book of Revelation as only a prediction of the future. Too many have done that and in it have created hellish doctrines. David Koresh and Jim

Jones were both fixated on the prophecy in Revelation and went off the deep end with it. Both ignored other major portions of scripture.

Over the years I have studied some of the good, the bad and the ugly. Jones and Koresh were among those who I spent time looking at. One of my novels, "The Handshue Sect – A Cult Gone Array" is a study of what can go wrong. It is the result of these studies. There are other cults, the KKK for example was really more of a cult than a political movement. It misused scripture to show its validity. Many of the anti-Semite and White Supremist groups also misuse scripture too.

God is not the author of confusion, he is not the purveyor of fear. He is not the author of hate. When someone is engaging in these, they are not preaching His word. There are cults out there that look good but are not.

In 1982 a set of audio tapes was produced and sold called, "One World Government 1983." It was billed as 100% bible based and sold a lot of copies. It created fear and confusion in the church. Initially my pastor was wound up on this, everyone had to hear it. I listened to the beginning of one of the tapes, I think there were 5, and stopped. I was young and not as well grounded in the faith as today, but I saw things that did not seem right. I didn't bother with listening further and after a short time the pastor started seeing holes in it and wisely withdrew his endorsement. As an aside, this is one of the men who have had a role of shaping my life in Christ. That he could be drawn away like this has done nothing but increase my concern that any of us can be subverted by Satan.

There are several things I look at when I see teachings and books. I look at the author. Can I find out what his track record is? I will more easily look at an unknown than a known with a tarnished record. I will look carefully but I will look.

I think ask another question. How big is the hype? The bigger and louder the hype, the less I pay attention to it. The devil uses hype, God does things in quietness and in order. I am a part of a Search and Rescue team. At searches, Dog teams show up, frequently some are self-dispatched. Confusion kills and a self-dispatched team often is one that brings confusion. That alone is a bad thing, like self-promotion of prophets. The next thing I look for the size of the K-9 sign and trappings on the vehicle. The bigger the sign, the bigger my concerns. I look at what results are coming from it, call it fruit. God is not the author of confusion and fear. If someone is using His word and creating that, they are polluting it.

After I get past the initial screening, I look inside. This set of tapes was suspect, I listened to part of one that was loaned to me and handed it back. I did not listen further.

Well, 1983 passed and no OWG. You would think the preacher who produced the tapes would have crawled in somewhere and hid. No way. He produced an additional tape, "One World Government 1984" that gave the explanation why it didn't happen in 1983 and it was sold with a note that if you did not have the 1983 tapes you needed to buy them to understand the 1984 tape.

Jesus told is to be cautious.

> Mat 10:16 Behold, I send you forth as sheep in the midst of wolves: be ye therefore wise as serpents, and harmless as doves.

There many who would mislead us, not all but many for profit. But the scripture is for our benefit.

Unfortunately, there are those who try to benefit from the scripture at the expense of the flock. Over the years I have seen shady characters in business, government and sadly in the church. There have been times I would have liked to take a whip and run them out of the church like Jesus did the money changers in the temple. And worse, I may have missed God's best for the church a couple times by not doing so.

But those in the temple and those in the church today who steal are not always the money changers. Although they stole money by cheating the people with the complicity of the priesthood the priests outdid them. Far worse than stealing finances is what is too often happening, stealing God's best from the people. The flock of God has the right to be led by a shepherd and under-shepherds who care for the flock, for love, not for money. Please understand me well. I am not like the congregation Oral Roberts talks about his dad preaching to. He characterized them, "Lord you keep him humble, we will keep him poor." He suffixed that remark, "And they did." I believe a workman, whether he makes tires, programs computers, writes legal briefs or leads a church is worthy of his hire. He is to be

14

properly compensated and if a church is not doing that, they are failing.

For some years Pastor George Mouer was a mold maker in a cast iron foundry while pastoring a church near Fayetteville. The church gave him some limited compensation. At one time the church could not have supported his family but as it grew his responsibility increased as well as his reach into the community. By the time he became full time the church was well able to support him. George passed away a few years ago but his legacy remains in the lives he touched.

Rev 2:2 I know your works and your labor and your patience, and how you cannot bear those who are evil. And you tried those pretending to be apostles, and are not, and have found them liars.

Here the church as Ephesus is being commended. "I know your works and your labor and your patience." How much more could someone desire to be commended for? Our works are not to achieve salvation, but we do works because we live for God. They are works OF righteousness, not works TO GAIN righteousness. We are to labor in His vineyard. Through faith and patience; we receive the promise.

He goes on with them not being able to bear those who are evil.

We find the sentence, 'Tried those pretending to be apostles." Peter talks of Satan and those who would do his evil works.

> 1Pe_5:8 Be sensible and vigilant, because your adversary the Devil walks about like a roaring lion, seeking someone he may devour;

He tells us to be vigilant because Satan is seeking who he may devour. One of his greatest tools are those who pretend to be of God and are not. We are not likely to be pulled away from the gospel by a Satanist or Wiccan. But we are more likely to be pulled away by someone with a tainted gospel that sounds good.

Paul expressed that concern to the Corinthians.

> 2Co_11:4 For if he that cometh preacheth another Jesus, whom we have not preached, or if ye receive another spirit,

15

which ye have not received, or another gospel, which ye have not accepted, ye might well bear with him.

He feared that if someone came to them who sounded good, preaching a false doctrine they might listen to him. Be careful with those who come speaking with words that sound good but carry a different message. Paul also reminds the Galatians of this.

Gal 1:8 But though we, or an angel from heaven, preach any other gospel unto you than that which we have preached unto you, let him be accursed.

Gal 1:9 As we said before, so say I now again, If any man preach any other gospel unto you than that ye have received, let him be accursed.

One of the greatest propogandists of all time was Steften Dalmer – a Brit who lead the black ops program in World War II. Unlike BBC that broadcast as BBC in German, this group pretended to be loyal German stations and fed to the German people a tainted news. Steften told the people who worked there, cover, cover, cover, dirt. The false stories had to be hidden, like the spoonful of sugar that makes the bitter medicine go down. As I read that book, Black Boomerang, back in the 1980's I saw how the church has been diverted by those who preach a doctrine that is cover, cover, cover, dirt.

As I study more, learn to know God's voice more, there are many in this world who at one time sounded good who do not now. I just recently on Facebook pointed out to a nationally known pastor that his teaching has deviated from both what he once taught and more significantly, from the pure word of God. He has allowed culture and his historic ethnic background to creep in. He did not give my post a 'like' but some others did. I have seen this before with other leaders. One of the problems with me, I have a darned good memory. If I am not sure I know where to research. I am not as much moved by what someone is hyping now. I see it in politicians, I see it in people and worse, I see it in the pulpit too often.

I recently found a video of Chuck Schumer talking about illegal immigration in 2006 and what he said, if it had been recorded with someone who could imitate Donald Trump's voice, would have been almost verbatim what Trump is saying and Schumer is taking him to task for it.

I recently was told, "Be more open minded." I looked at the person who at one time would have called abortion wrong, called homosexuality not God's plan and shook my head. They were asking me to accept these as good. I responded, "If we get too open minded our brains will leak out."

Yes, I took it on with those who would pervert God's word. I was called of God to be a watchman. If the watchman blows an uncertain trumpet, who will prepare for battle. Who will wake up and meet the enemy or who will be killed as they sleep? No, I do not wish to offend but I must be true to God. And I can tell you that when one is awakened from their sleep, both physical and spiritual, they will not be cozy with the watchman. But I must be, like Paul, obedient to the heavenly vision. Yes, it would be easier to let them sleep but then their blood would be on my hands.

As an aside, my response to those who would be involved in homosexuality, abortion or anything that God has defined as outside his will is to treat them as publicans and sinners. That response is to love them into the kingdom, not beat them over the head. I cannot condone, legitimize or participate in their sin. But I can love them. I can work to bring them to a better understanding of God. But then, I can say the same for the drug addict, the adulterer or the robber.

Rev 2:3 And you have borne, and have patience, and for My name's sake you have labored and have not fainted.

I am sure that if any of the church at Ephesus were to hear this they would be smiling. I am sure those in my church would be smiling if god said that about them. They would be saying to each other, "See. We are doing it right." I have a question, "Are you?" Have you looked? Are you measuring the works of your body against God's design or the other churches, being the real cream or being the cream of the crap? God does not mark on a curve.

Have you taken out the plumb line of the word of God and held it up to measure your works? I have used a plumb line to check a tower by just holding it up and looking past it at the tower. It works. No, don't sent me an e-mail about how I have offended you. Get out the Monday and true the columns of your life.

What is a Monday? In high steel construction is a very heavy sledge hammer. Historically when they were building a tall building, they used a plumb line to check the beams were vertical on one floor, then verified that several floors were still vertical. The only limitation on the plumb line was the length of the string. If the columns weren't true, perfectly vertical, a worker would take a Monday and hammer them true. It is hard work. Know why the hammer was called a Monday? Because when you used one for a while it didn't matter what day of the week it was, you felt like it was Monday Morning! Sometimes when I am working to straighten my life it feels like Monday morning no matter what day of the week it is. Sometimes I need some real hammering. If you think you do not, I can assure you, you do.

We need to labor and not faint, to spread the gospel but more important, align our lives with His plumb line. Without that, we will never be effective in his service.

God is not a respecter of persons. Look at what the Apostle Paul said.

> 2Ti 4:7 I have fought a good fight, I have finished my course, I have kept the faith:

> 2Ti 4:8 Henceforth there is laid up for me a crown of righteousness, which the Lord, the righteous judge, shall give me at that day: and not to me only, but unto all them also that love his appearing.

This is what the ones who remain faithful to the calling on their lives can expect.

I have joked at times that we need to change the words of the song, "When the saints go marching in"[3] to "When the saints come dragging in." I am not talking about the physical, the changes in our bodies as we grow older. I am talking about our lack of zeal and enthusiasm for the things of God. Those of us who are older, whose knees are getting a little shaky need to be the ones who show the younger the way to walk with God. We may not be able to run the

---

[3] The origin of this song is unknown. I believe it has great lyrics but has been tainted by the world's use of it. I still would love to hear the church use it. There are fantastic lines like, "When they crown Him Lord of all."

marathons, but we can walk that walk, even if we need a cane or walker. We can even walk that walk from a wheelchair.

Rev 2:4 But I have against you that you left your first love.

This was one of the scriptures that several of the "one scripture preachers" used frequently used, in fact in nearly every sermon. My dad called them that because no matter what the message was about, they somehow got back to this or the 'neither hot or cold" verse. There are 31,201 verses in the King James version of the Bible. All scripture is profitable, what about the other 31,200.

I struggle a little with the abrupt change in this verse. The previous says so much good. What could this church have missed? I have seen many explanations of the 'left your first love.' I believe it could be a combination of the love for Christ, the love for one another, the love for God, the love for the sinner. I believe the previous verses hit at the problem. Although they look so good, there is some self-righteousness buried in it, the works, the labor, the patience, what they were doing. There may have been some pride buried in it. "We are doing it." They may have lost the need to lean on Jesus. I wish I could tell you I have not seen this in churches.

I see this at times in churches that have had a good time of building in numbers. They start to get an attitude that they have it all right, others should look to them and see how great they are. They begin to sing, "How Great We Are", not "How Great Thou Art."[4] It never comes out that bluntly, but it is there. I will allow the word of our Lord to help you.

> Mat 7:22 Many will say to me in that day, Lord, Lord, have we not prophesied in thy name? and in thy name have cast out devils? and in thy name done many wonderful works?

> Mat 7:23 And then will I profess unto them, I never knew you: depart from me, ye that work iniquity.

I have been in two churches that were growing both in numbers and spiritually. The one in Illinois was one that pretty much if not totally avoided this. Trinity Mennonite on Fourth and Queen Streets in Morton Illinois was one of, if not the leader church in the Charismatic Mennonite movement. But I look at the men and

---

[4] Awesome song by Bill and Gloria Gather.

women there, the Miller's, the Eigsti's, the Kellum's, the Hoffer's, and some more I can't remember names, the attitude of humility, service, and knowing they were dependent on the Holy Spirit and what transpired there does not surprise me. In seven years, they grew from a meeting of 35 in the yard of a house on Fourth Street to over 375. They never tried to build a big church, they just tried to serve God and the people in their community and it grew.

I know some of you are thinking. He was in Pentecostal tent meetings. They were weird. Now he tells us he was a Charismatic Mennonite. It is getting weirder. What's next? As Todd says, "Did you worship with snakes?" Answer, "No. Never. Will not do."

Let me tell you about Trinity. They reached out to other churches whether they embraced the charismatic or not. They were committed to following God, but they worked hard to not offend. They were salt and light to other churches without having the air of superiority, something I never saw or even suspected. They held services at times with the other Mennonite churches in the area and were very careful to not offend. There were three Mennonite churches in Morton, the very Traditional First Mennonite, Trinity the Charismatic and Fourth Street Mennonite, the one in the middle. Their modes of worship were different but their mode of caring for each other and the community (meaning the world) were very close to identical and they worked together. I never saw compromise of the faith, but I saw a group of people (really three groups) who loved God, loved people and wanted the best for everyone. They shared their faith freely but were careful to not offend or be condescending. Although I am sure there were at times some minor transgressions, I never saw anything that concerned me. I wish I could have said the same about some other groups I have worked with. I have seen some who seemed to enjoy offending. I am reminded that you can catch more flies with honey then with vinegar. I am also cognizant you cannot compromise the faith.

The other church I was in that was in a similar growth situation but after a while got to the air of superiority and displayed it well. It was not long till things started to go wrong. When elitism sets in it is like a virulent infection. It grows in intensity and damage. The focus moved to dynamic worship, not on God. As an aside, I am not sure what dynamic worship means to them but to me it means meeting and honoring God. That can be done at sound levels between 30 db (quiet) and 100 db (very loud). To the musicians who read this, call it below ppp and above fff. There was a series of worship leaders and strife was rampant. People were used and discarded. The pastorate became a royalty. The worship team was nearly always a source of contention. Anyone who did not come up to the standards was not welcome. The church growth had been based on His word, meeting people where they were and helping them grow. This was snuffed out in the fray and died. They wanted "professionals" in every position. And it radiated out to impact several other churches. Internally, first stagnation set in, then decline. To a church group I mentioned one day that professionals built the Titanic and amateurs, those who did it for love, built the ark. Both sailed for about 40 days, one completed its voyage, the other made it half away. I got looked upon as if I had pulled on Superman's cape. I may mention that a person can be both a professional (paid) and be an amateur (doing it for the love of it) at the same time. Those who do it for the love of doing it are the ones who in business, the church and government stand out as targets of those who will do it for gain, even if that gain is prestige and power because they are threats. They are the cream that tends to rise without trying. They are also the ones who carry most of the load and are the most productive.

Clearly that body had lost their first love. There were ones in the body who saw this and were trying to pull the boat back on the right course, but they were at first ignored and tolerated. That quickly changed to them being shuttered off and finally some were encouraged to leave. Three weeks after we left – with God directing it – we had not attempted to contact any members, we were mentioned from the pulpit and the pastor told the congregation it would be best if they had no contact with us. It was just short of shunning. We stayed clear, we had no desire to tear down what we had helped to build.

Least one of you ask, why did you leave the Mennonite congregation you talked about? A job situation moved us 800 miles to Morton and then three years back to York. We could no longer be viable members of that congregation. It was not our fault or theirs. I believe God had a plan in our moving there and our leaving. We grew and we contributed to that body. The company was the vehicle he used to make that happen.

I study fruit, not grapes and strawberries, but the fruit of the actions. That church that lost focus finally stepped across the line. After one terrible night when many in leadership were removed for reasons that were just plain wrong and unscriptural, the church took a serious decline. I was in that meeting and a short time before the firings, I read the Revelation 2:1-4 aloud to the group. God placed it in my heart, I had no idea why, but I knew it had to be read. The pastor had lost the first love, for God, the people and Christ. He was working on an ego trip, to build a mega-church.

What about the fruit? The church has built a new building in another location. In 21 years, there have been two subsequent pastors. The previous ones are not in ministry. The church has never approached the size of a mega-church and is to some degree, struggling with finances. I am as convinced today as I was then that we should aspire to grow God's Kingdom, not ours. If ours grows in the process, so be it. God is committed to support His plan, not ours. If our plan lines up with His, even if it is close, it will be made to happen.

Rev 2:5 Therefore remember from where you have fallen, and repent, and do the first works, or else I will come to you quickly and will remove your lampstand out of its place unless you repent.

This is an awful indictment for a church that has heard the accolades of the first verses. Look at where you were. Repent. Do the first works. This is a pattern for any of us, personally or as a church if we have missed it. Look at where we were when God's power and presence were real to us. I will discuss Sampson elsewhere but when he becomes humble the scripture says that the hair of his head began to grow again. It was then he started back. As individuals and churches; we need to do this frequently. I am not sure what the interval should be, but it definitely should not be decades. I do not advocate we sit and say, 'woe is me'. It is to look at where we could be and if we have misses it, do what is needed to return to that level of glory and just as important, grow beyond it. We need to be always pressing to follow God more closely.

God has no desire to remove our lampstand. He will restore us if we will just return to Him.

Rev 2:6 But you have this, that you hate the deeds of the Nicolaitans, which I also hate.

Even with what negative has been pronounced comes this. They are working to hold to the faith, in a place where others have failed and followed false teachings. They are viewing the false teaching as wrong. Note here the is it not that they hate the Nicolaitans, but that they hate their deeds. This is a critical distinction. We must never hate those who have missed the mark. We may and should hate their deeds. But for them we must have love. We must treat them as publicans and sinners, we work to bring them back.

I recently read a proposal for mainstreaming LGBTQTG into a mainstream denomination. I hate the deeds, repeat deeds, of the LGBTQTG. I will say without any equivocation, the thrust of this group is not Godly. Let me be clear, I do not hate them, I feel sorry for them because most if not all of them are miserable. There can never be in me a hatred for any one of them no matter what they do and I find challenge here since there are times they harm others and it would be easy to do so. This act I hate but they are God's children, worthy of my love and concern. I will both work to bring them to an understanding of God and to repentance and ultimately forgiveness. I will not participate on or abet their sinful ways.

About 25 years ago a gay group in York was holding meetings in a local church and was inviting early teen boys to their meetings to 'explore their sexuality.' We learned this because we were part of the community access TV and it was mentioned on one of their programs!

We quietly apprised the church. I am not sure if our action brought the change, but the meeting place was withdrawn by a very liberal church because they saw concern about culpability if something went wrong. There was no hatred for this group, but they crossed a line of propriety and one of our officers took an action. I have a way of testing my bias on this. I asked, "What if heterosexual guys over 21 were inviting 15 year old girls to a meeting to explore their sexuality?" I had my answer. I would have easily opposed it.

In this manner of determining bias, I follow the scripture.

23

Eze_45:10 Ye shall have just balances, and a just ephah, and a just bath.

Act_10:34 Then Peter opened his mouth, and said, Of a truth I perceive that God is no respecter of persons:

Jas_2:3 And ye have respect to him that weareth the gay clothing, and say unto him, Sit thou here in a good place; and say to the poor, Stand thou there, or sit here under my footstool:

I cannot claim to follow God and be a respecter of persons, i.e. be one who practices discrimination. I lived that in the business world, giving opportunities to worthy females and minorities and even at times taking a chance on someone I could see could grow into the job. Looking back, I am not sure I didn't miss some worthy ones because I was never disappointed in the person I selected. If I was right that often I may have been too selective. But if I was, it was because the person did not exhibit the qualities I sought, or my criterion was too strict, not the gender or the ethnicity.

I hate the deeds of any who will work to harm others, be it spiritually, physically or mentally. I do not hate them however I will work to make it more difficult to do harm or better, if possible, make it impossible to continue their evil.

Rev 2:7 He who has an ear, let him hear what the Spirit says to the churches. To him who overcomes I will give to eat of the Tree of Life, which is in the midst of the paradise of God.

I may have something in common with the writer here. I ramble and in this he appears to depart from the theme. It is a call to hear the word, hear what the spirit is saying. And it is a promise. To him who overcomes I will give to eat of the tree of life. And how do we overcome. Our own works? Our strength? Keeping the Law? Observing Lent? Saying Hail Mary's? Later in the book John writes the answer.

Rev_12:11 And they overcame him because of the blood of the Lamb, and because of the word of their testimony. And they did not love their soul until death.

These were the ones who overcame through the worst of times. It was by the blood of the Lamb and by the word of their testimony.

The songwriter asked, "What can wash away my sin?[5] What can make me whole again?" And he answers that question each time. "Nothing but the blood of Jesus." Only this can be our cleansing. And it can be such only if we appropriate that sacrifice to our lives. It can only be if we testify that.

I am reminded of a song, "Let me loose myself and find it Lord in Thee."[6] It begins with verse one.

"Many years I longed for rest,
Perfect peace with in my breast,
And I often sought the lord alone in tears.
But I would not pay the price,
Would not make the sacrifice,
So I wondered on and on for many years."

It looks almost hopeless but the writer steps through a verse where he says that he hears Jesus and although the step was hard, he

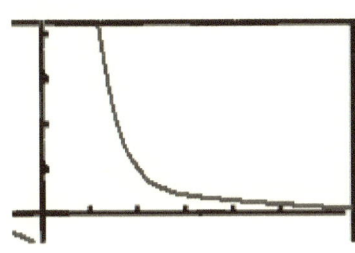

followed. The last verse is my testimony. No, I am not there, I have not arrived, but I am moving in that direction. I am like the asymptote, a line defined by a math expression that approaches a line but as the value of x increases, never gets to it.

I am approaching something I will never reach till I cross that river.

Now the blood has been applied, [7]
By his power I'm sanctified,
And the savior brings me constant victory,
Yes, he took away my sin,
Washed and made me pure within,

[5] Nothing but the Blood   Robert Lowry 1876
[6] Copyright 1943 Renewed 1971 Ross H. Minkler. Assigned Singspiration Music (Admin. by Brentwood-Benson Music Publishing, Inc., 741 Cool Springs Blvd., Franklin TN 37067)
[7] Ibid.

And I lost my self and found it Lord in thee.

I have been asked why I so frequently quote scripture. Jesus said often, "It is written." If it was good enough for Him, it is good enough for me. I also get asked why I quote the songwriters. Many of their works are time-tested, have stood the tests of time and are good doctrine, they follow scripture. I will add, there are some that are not. Many of them were catchy, were there for a time and faded. I find that writers like Ira Sankey, Ira Stamphill, Fanny Crosby, Stuart Hamblen, and even some more recent ones like David Engels and David Baroni stick to the concepts of scripture. Many of them talk to God, to show our love for him. Many speak to our peers, telling of the love of God, encouraging the faithful, and calling to those who don't know him. Some like "It is well with my soul", an awesome piece, bridge all of these. But no matter what, any that portray the truth, are profitable.

I mentioned "It is well with my soul."[8] The third verse, the most rarely sung, bridges these masterfully.

My sin oh the bliss of that glorious thought.
My sin not in part but the whole.
Has been nailed to the cross and I bear it no more
Praise the Lord, praise the Lord Oh my soul.

I had heard that song many times and then one night I played a Lynda Randell version and she hit that verse. "My sin not in part but the whole. Has been nailed to the cross and I bear it no more." This is the work of Christ on the cross. All my sin, not just part of it was nailed to that cross. It is a reminder to each of us, that our sin can be forgiven, not just some of it but all of it, and we bear it not more. We are free. It is a reminder to each of us, saint or sinner. And the last line reaches to God, "Praise the Lord oh my soul." It places the thanks where it belongs.

In this sign we conqueror, the cross. In it we are victorious. Through it, and only through it we will have the right to eat of the tree of life.

---

[8] By Horatio Spafford and composed by Philip Bliss. First published in Gospel Songs No. 2 by Ira Sankey and Bliss (1876)

We have triumphed through the blood of Jesus.

I remind my friends. God's retirement plan is a home in heaven not a rocking chair here in a rest home. I would like to have us see the mature saints as I do the Army Rangers and the Navy Seals, people who have trained to be the best they can be and are experienced in battle. They should be the ones who lead us into battle, direct our walks, train up and mature the younger ones. There is not a place for the rocking chair unless it is used as a place to teach, lead and pray from. I am seventy five. I have retired from the workplace. I have not retired from serving Him and people around me.

SMYRNA

Rev 2:8 And to the angel of the church in Smyrna write: The First and the Last, who became dead and lived, says these things:

Christ says to us, I am the alpha and omega, I am the first and the last. I have been, I am, and I will be from eternity past to eternity future. I am he who became dead and lived. Not killed but became dead. His life was not taken, he gave it up. He has a right to speak into our lives. We have the responsibility to listen. In this we see the work of Him on the cross! He was dead and buried. He is not there in the grave, he has risen. He has conquered death, hell and the grave. Because He lives, we too will live for evermore.

The leaders of other religions are still in the grave. Mohammed is buried at Medina. Buddha's ashes were scattered. Jim Jones and David Koresh are still in their graves. But the tomb of Jesus on Nazareth is empty. He has risen. He has risen indeed.

Rev 2:9 I know your works and tribulation and poverty (but you are rich), and I know the blasphemy of those saying themselves to be Jews, and are not, but are the synagogue of Satan.

I know your works. Smyrna has been known to do good works. That can be said for many churches. I know your tribulation. I know the trials that you have gone through. And I know how you have been poor, but I must remind you, you are rich. You are rich not because of physical riches but because you have my life and power that you have appropriated. We must accept that power and life, it is not automatic.

They have withstood the blasphemy of those who are doing the work of Satan. These who opposed them were not wiccans or Satanists. These were professing to be Jews and were not. They carried a doctrine that was false. Let me address something, what constitutes false doctrine? I can tell you it is not a different way of kneeling, a mode of worship, the things that all too many key on. It is any doctrine that does not speak of the incarnate, crucified, risen and ascended Christ, the Son of God. That includes Satanism, sure, but it also includes many other doctrines that look and with some perfume of deception smell more respectable. In God's plan, respectable doesn't cut it. You can put perfume on a pig, but it is still a pig. You can 'dress up' a false doctrine or perfume it to make it smell better but nothing can make it other than what it is, a false doctrine. The Apostle Paul sets the standard in these verses.

> 1Co_12:3 Wherefore I give you to understand, that no man speaking by the Spirit of God calleth Jesus accursed: and that no man can say that Jesus is the Lord, but by the Holy Ghost.

> Gal_1:8 But though we, or an angel from heaven, preach any other gospel unto you than that which we have preached unto you, let him be accursed.

> Gal_1:9 As we said before, so say I now again, If any man preach any other gospel unto you than that ye have received, let him be accursed.

Paul tells those at both Corinth and Galatia in no uncertain terms, this is the gospel and there is no room to deviate from it. I can point to something here to help you see false doctrine for what it is. When someone changes Jesus from the sacrifice for our sins, from the one who was there in the beginning, from being the only begotten son of God, they are polluting the gospel and I will say as Paul, have nothing to do with them. I will also say, if you have God's word to move, stand against them, refute them and call them to come to a pure knowledge of Christ.

There is one group that says Jesus had a brother, Lucifer. The Muslims call Jesus a prophet, but they deny him as the incarnate son of God and savior. They call Christians polytheists, having many

Gods because we talk of the Father, Son and Holy Spirit. Let me be frank, they do it out of ignorance because they do not understand that there is a mystery here on how three can be one.

I mentioned Rev. George Mouer elsewhere. He was for some time a part time pastor who worked at a Foundry in Waynesboro Pa, one of the dirtiest places. He told a story about his early years as a mold maker. They use a sand to make a mold and then pour molten iron in and let if cool to make an item. The sand is broken away and you have the item. The sand mold is very fragile. George tells how when he was starting out he made a mold, it wasn't right and he tried to patch it. One of the senior mold makers looked at it and told him to dump the sand and start over. He told George, if it isn't right you dump it. When a doctrine isn't right, you dump it. George was great at using the natural to teach God's word but so was Jesus.

And Paul so desperately wanted to the Galatians to hear it that he writes it twice in that letter. When someone repeats something, they consider this important. And in the third chapter to the Galatians he warns of diluting or adulterating the truth.

> Gal 3:1 O foolish Galatians, who hath bewitched you, that ye should not obey the truth, before whose eyes Jesus Christ hath been evidently set forth, crucified among you?

> Gal 3:2 This only would I learn of you, Received ye the Spirit by the works of the law, or by the hearing of faith?

It clearly shows that the false doctrine they were heading back to was justification by keeping the law, and looking back one verse, it is those who were teaching this who he was targeting. He was not targeting card carrying Satanists, he was targeting Torah carrying ones who were teaching a return to the salvation of Judaism! It was what Paul was preaching only a few years before and was willing to commit murder to promote. But Paul had an experience on the Damascus Road and he could not deny what happened. Like the songwriter who said, "I was there when it happened and I guess I ought to know!"[9] I am sure Paul also knew how enticing the comfort of the law was. You didn't have to think to follow the law. The words were written on tablets of stone and on scrolls as the Jewish leaders added to the law. You could point to a place on the scroll

---

[9] Words and Music by Herbert J. Lacey, 1920

and say, "That is the law." Today we have the more general law, "Love the Lord your God with all your heart and your neighbor as yourself." We don't have an extensive lawbook to check if this is violates a section or sub paragraph, but we have the Holy Spirit to guide us, however to the Jews, this change in paradigms had to be daunting. We can be disparaging but let's be honest, this was a difficult change and it was easy to go back to what was comfortable. Understand, like Paul, I will emphatically say, "wrong." But I see Christians today making the same error.

> 2Co 3:6 Who also hath made us able ministers of the new testament; not of the letter, but of the spirit: for the letter killeth, but the spirit giveth life.

We must look to the spirit of the law, not the letter. Allow me to give you an example. I have heard Christians complain about and rationalize about breaking speed limits. Although I see some of them as improperly set, I have another issue here. Note, I do all too often without malice step across the line. But if I care about my fellow man, can I legitimately break those? And even if I can justify that, how can I justify breaking any civil law?

> Rom 13:3 For rulers are not a terror to good works, but to the evil. Wilt thou then not be afraid of the power? do that which is good, and thou shalt have praise of the same:

> Rom 13:4 For he is the minister of God to thee for good. But if thou do that which is evil, be afraid; for he beareth not the sword in vain: for he is the minister of God, a revenger to execute wrath upon him that doeth evil.

> Rom 13:5 Wherefore ye must needs be subject, not only for wrath, but also for conscience sake.

Now that gets down to where the rubber meets the road. I have to obey that law, not just because the officer carries a weapon or a ticket book, I have to obey to have a clear conscience. I can tell you, that is a challenge for me too.

Rev 2:10 Do not at all fear what you are about to suffer. Behold, the Devil will cast some of you into prison, so that you may be tried. And you will have tribulation ten days. Be faithful to death, and I will give you the crown of life.

This verse would seem one of gloom. Most of the church today would pull back if they heard it. It pronounces some very certain suffering. And it gets worse. Some will be imprisoned. They will have tribulation for ten days. It implies some will die. But it ends with the hope, be faithful and He will give a crown of life. It is in line with the scripture;

> Mat_10:28 And fear not them which kill the body, but are not able to kill the soul: but rather fear him which is able to destroy both soul and body in hell.

If we are to be faithful to Him, we will have to die, all of us except those who are alive to see his appearing. Most will not face the death of a martyr, but if we aspire to live for Him, we will all have to die to self. I mentioned a song earlier, "Let me loose myself and find it Lord in Thee."[10] The verse I mentioned ended negatively, "then I wondered on and on for many years." I did not mention this verse then. It seems appropriate now.

> Then one day while knelt in prayer,
> Jesus whispered to me there,
> "Take your Cross and follow Me to Calvary."
> Oh, how sad it was to die, and all self to crucify,
> Just to lose myself and find it, Lord, in Thee.

There must be a denying of self. In a sense it is a death to self. It may be a traumatic, all in one moment as the martyrs of the past and present, or it may be a process over time. For those who do not know, there are people being killed for their faith every day. In some areas of the world holding to Christianity can bring a death sentence, particularly in China and in Muslim countries. The Muslim apologists will tell you that is Christian lies. Others will tell us it happened in the past, they don't do it today. There have been beheadings in the last two years and although they appeared to be political, they were focused on having the person embrace Islam. I challenge you to read the news stories, not just the headlines. When you read the whole story you get, more of the story, not just what the editor wants you to focus on. Often the truth leaks through.

---

[10] Ibid

31

But death to self must happen, at some point we must die to self if we are to be victorious in Christ. The apostle Paul writes:

> 1Co 15:31 Day by day I die, by your rejoicing, which I have in Christ Jesus our Lord.

It may be traumatic and I am pretty sure for those who have the dramatic event it is. That is the turn around point. The fine course corrections come daily. It is very interesting that Paul, the one who had the Damascus Road experience, certainly an epiphany, says he dies daily. Many of the disciples had significant experiences when Jesus said to them, "Follow me." Would that we would all not only hear but also heed that call.

But look at the last line, be faithful to me till death and I will give you a crown of life. Those who are faithful will live eternally in a place with Him. But it is if we are faithful to Him to death. I somehow see this as a verse with a double meaning. What does, "till death" really mean. Does it mean till we are martyred? An emphatic, "yes." Does it mean till we have spent our years here and are called home by natural death? Just as emphatic a yes. Does it mean till He returns and takes with Him the remnant, the glorious church without spot or wrinkle? Yes, emphatically, yes.

> 1Th 4:15 For this we say unto you by the word of the Lord, that we which are alive and remain unto the coming of the Lord shall not prevent them which are asleep.

> 1Th 4:16 For the Lord himself shall descend from heaven with a shout, with the voice of the archangel, and with the trump of God: and the dead in Christ shall rise first:

> 1Th 4:17 Then we which are alive and remain shall be caught up together with them in the clouds, to meet the Lord in the air: and so shall we ever be with the Lord.

No matter if we go by the grave or meet Him in the sky, we will all be with Him. Will that not be a wonderful reunion? There are not a handful I want to meet there, it is probably a couple hundred. One of the Camp meeting songs was Heaven (Happy Home Above).[11] I am posting it here because I want you to see its message. It

---

[11] Song Heaven Happy Home Above – Words and Music by Boyd and Helen McSpadden

embodies the role of the Christian, to meet Christ, invite Him into our lives and then work to being others to Him.

In childhood I heard of a heaven
I wondered if it could be true
That there were sweet mansions eternal
Somewhere up there beyond the blue
I wondered if people really go there
Then one day sweet Jesus came in
And I had a vision of heaven
My soul through all heaven I'll spend

Heaven (happy home above)
Heaven (land of peace and love)
Oh, it makes me feel like traveling on
Heaven (supernal), Heaven (eternal)
I'm so glad it's real

Then I got a vision of heaven
My soul overflowing with love
My heart like the Savior's is broken
For friends that will miss that home above
Then a voice from the hills of Judea
Still ringing words of sweet belief
Worlds of attraction don't thrill me
My soul has a change of relief

When we get a real vision of Christ, of his work, we will experience heartbreak for those who will miss heaven. When you get to that point you will no longer be damning the sinner, you will be praying for him or her, living a life to show them the way. I look at the next to last line about the attractions of this world. I see the most committed to Christ and I see the fruits of their lives. They are having fun in this life, but their greatest joys are in serving others.

I have been in the pulpit a few times. In one teaching I was talking about doing what we can to stop those who are on that road to hell. Yes, if they have not accepted Christ, they are on their way to hell. We need to get that picture. Hell fire and brimstone sermons should

33

be mandatory at least once a year, for the saints, not the sinners, to remind them why we are to present the gospel. While I was preparing one sermon, in my mind I saw a person on the road with a red flag, trying to stop traffic where a bridge was out. I was going to get a piece of red cloth and considered it theatrics until during the early part of the service. I pulled Connie aside, asked if the expressive worship team had some red material and a few minutes later she handed me about a two foot square of red cloth. I put it in my pocket, not sure if I would use it. I don't like theatrics. When I hit that point where I was talking about that missing bridge, I pulled it out and waved it. I told them. If you were on that road with that bridge out, you would get the biggest piece of red you could find, the brightest light, and wave it to try to stop the cars. We need to get the biggest piece of red, the brightest light we can find and wave it to try to stop those who are headed to hell. I pray that message was not missed by the congregation. As an aside, that was over 20 years ago, a few years ago I pulled out a suit I had not worn for some time, the red flag was in the pocket.

Let's get out that red flag. Let's wave it.

Rev 2:11 He who has an ear, let him hear what the Spirit says to the churches. He who overcomes will not be hurt by the second death.

This is a reminder for us to hear what is said. To me it is an admonition to take a careful look at what he has said to the churches, benefit from the encouragement and heed the warnings that are here. We should not pick and chose what we think but hear it all.

The last line is to every Christian. We are overcomers. The second death is not for us.

Oral Roberts preached a sermon, "You can't go under if you are going over." It was tied to the disciples in the boat with Jesus. When the storm hit, they thought they were going under. But what did Jesus say to them just before they pushed off?

> Mar 4:35 And the same day, when the even was come, he saith unto them, Let us pass over unto the other side.

Oral keyed on that line, "Let us pass over." If Jesus said, "Let us go over," you can't go under. We each need to make the promises he gave real to each of us. When they become real, we will walk in victory. We will be the head and not the tail. We will triumph, not in our strength, but in His.

PERGAMOS

Rev 2:12 And to the angel of the church in Pergamos write: He who has the sharp sword with two edges says these things.

The two-edged sword is used several places in scripture. The writer of Hebrews uses the reference to describe the Word of God.

> Heb_4:12 For the word of God is quick, and powerful, and sharper than any twoedged sword, piercing even to the dividing asunder of soul and spirit, and of the joints and marrow, and is a discerner of the thoughts and intents of the heart.

The Word is the discerner of man's thoughts and intents. It shows our thoughts and intents in the light of the word of God. It is ultimately up to us to use that to decide out course of action. We are to follow Him. Our lives must line up with Him.

The reference in Amos to a plumb line is a corollary to the two-edged sword. Technology has given us many new tools to measure and align portions of buildings however, many of them are subject to being damaged and must be calibrated. The one exception is the ancient plumb line, when held up, it is true. It is how buildings have been built for centuries. If the item, building, column, tower lines up with the string, it is vertical, the word used is in the building trades is "plumb". I have, when tools were not available taken a piece of string and a stone! If you are careful to select a stone that is somewhat symmetrical the plumb line will be true. Like the word of God, it will require no adjustment, no offset, no calibration. When we 'tweak' the word to fit our ideas, we contaminate it and the message of hope and love it would bring. We now have lasers but many of our large skyscrapers were built using a plumb line to hold it 'true', the word used to say the columns are vertical. Even today they are used in some places. I talked earlier about this in another sense. If we build out lives using the plumb line of the word, they will stand.

> Amo_7:7 Thus he shewed me: and, behold, the Lord stood upon a wall made by a plumbline, with a plumbline in his hand.

> Amo_7:8 And the LORD said unto me, Amos, what seest thou? And I said, A plumbline. Then said the Lord, Behold, I will set a plumbline in the midst of my people Israel: I will not again pass by them any more:

When we allow the word to become a part of our lives, allow it to permeate our thoughts and actions, our lives will align with Him. The two edged sword of the word is both the discerner and the standard.

Rev 2:13 I know your works, and where you live, even where Satan's seat is . And you hold fast My name and have not denied My faith, even in those days in which Antipas was My faithful martyr, who was slain among you, where Satan dwells.

This sounds like a near perfect church. The church at Pergamos was facing a terrible world. It had gotten to the point that the opposition was willing to take lives. We, in the US are somewhat protected compared to the Coptic Christians in Egypt, Christians in Muslim lands, and some in China. There were and still other places where being a Christian is risky. A Coptic Christian

IN OUR great land, where freedom of religious expression is supposed to be an honorable pastime, something very sad took place last week. A group of Muslim-American clerics and community leaders felt the need to call a press conference in Jersey City to declare that members of their faith were not ruthless killers.

It was a pathetic moment. But it was necessary nonetheless after a week of continuing rumors that the brutal murders of four members of an Egyptian-American family in Jersey City was the work of Muslim extremists – indeed, that the act was a ritualistic execution of Christian "infidels" by Islamic enforcers.

Police were quick to say that there was no credible evidence to back up such rumors. But it did not matter. When you mix a real murder with centuries-old fears of religious extremism rooted in the old ways of an ancient Middle East homeland, strange things can happen on an American street.

MIKE KELLY

RECORD COLUMNIST

Consider the scene outside an Egyptian Coptic Church on Jersey City's Bergen Avenue last Sunday. Two days earlier in a nearby home, police discovered the bodies of church members — Hossam Armanious, 47, his 37-year-old wife, Amal Garas, and their daughters, Sylvia, 15, and Monica, 8. By Sunday night, hundreds of Egyptian-American Copts stood outside the church while members of the American Coptic Association announced that the murders were really an "execution" similar to those by anti-American terrorists in Iraq.

"Wake up America!" declared the association's president, Dr. Monir Dawoud.

And so, the speculation swirled.

That morning, newspapers carried stories that the victims, all devout Coptic Christians, had been tied up, and gagged, with their throats slit. Those same news stories also said that police and FBI agents were checking a tip that the family may have been targeted for death by Muslim extremists after Hossam Armanious participated in heated religious debates on an Internet chat room.

family was killed in New Jersey a few years ago. [12] These, and some others could be faced with giving their lives for the faith. Many of the groups that record hate crimes ignore any hate cries against white protestants. But even here in the US, there is some persecution. The baker who refused to make a cake for a gay wedding certainly saw persecution. Crosses placed along roads have been targets of legal action. Even the ten commandments have been under fire.

Although I am not sure that refusal to bake a cake was the right hill to try to hold, the one to die on, I am not critical of him, I know all too well I may be faced with 'holding some undefendable hill' at some time. It is apparent that the gay couple picked the venue, the state, the baker to create an incident. They deserve nothing but contempt.

The Supreme Court of the US has made some terrible rulings against life and religious freedom. Many, if not all these rulings have been modifying the constitution. The current court has a better record with the most recent appointment and another is in the wings as I write. [13]These two men are ones who believe the constitution is the law, they have no right to change it.

Rev 2:14 But I have a few things against you, because you have there those who hold the teachings of Balaam, who taught Balak to cast a stumbling-block before the sons of Israel, to eat things sacrificed to idols and to commit fornication.

This verse and the one that follows shows us something when compared to the one before. We can be doing so much right, and yet we fail in others. One of the stenches to the church over the years is this 'cafeteria Christianity' the taking what we like and ignoring other points. I have seen churches that preached against bingo, smoking, drinking, and chewing but tacitly accepted adultery, even in the clergy! I will be frank here, what I am talking about here is not speculation, gossip or the like, I have concrete facts and could

---

[12] The Record (Hackensack, New Jersey) · 23 Jan 2005, Sun · Page O1
[13] He has since been confirmed. As I write this Ruth Bader Ginsburg has had surgery for cancer. If she resigns, we may be able to put another good justice on the court. I am not looking to make the court Christian, the role of the court is to interpret the law, but this change will make it constructionist, upholding the constitution. If that is done we will have religious freedom.

name names if there were reason to do so. I want to be clear here, I do not see preaching against bingo, smoking, drinking, and chewing as sin when handled in moderation. On the other hand, a little adultery is like a little leaven, it impacts the whole loaf. I will comment here, most clergy is honest, trying to live right and doing very well, but there are some who do not seem to have the right relationship with God. We have seen it with the televangelist scams but unfortunately it is not limited to that subset.

I am going to dive deep here. I know Jim and Tammy Bakker did wrong. But there are others I blame in that scandal. Some of that blame must be laid at the feet of those who worshipped them. This is one way to destroy someone, lift them up till they fall victim of pride. The second group I blame is those who were close in, the ones surrounding them. There had to be at least a dozen in the ministry who knew what was happening, what was going wrong, and they did not care enough about their brother or sister to say, "Jim, Tammy, this is wrong. You can't do it." It would have been done at the risk of their job, but love demands risks at times.

This happens in the entertainment business, the ones who have died of overdoses of drugs for example, there had to be some around them who knew the risks, helped them acquire the drugs and some who didn't do anything to stop it. The pressure on these to perform must be intense, many are terribly insecure, they 'need' drugs to keep going. Many performers look confident on stage but are really scared little boys and girls. The promoters are willing to help them get them to keep making money even if it kills the goose who laid the golden eggs. What has been done to those like Elvis, Michael Jackson, and a host of others is criminal, people should be in jail. The charge should be depraved indifference or maybe murder. I ask the question, "Am I my brother's keeper?" I know the answer, "Yes." It is apparently not clear to those who destroy lives.

Rev 2:15 So you also have those who hold the teaching of the Nicolaitans, which thing I hate.

I will not get into the discussion of what the 'teaching of the Nicolaitans' means other than it was not pleasing to God. What else do I need to know about it? That is bad enough. However, I can very well define in by the two references here, the likening it to the teaching of Balaam and the statement, 'which thing I hate'. We can define it as a false doctrine with either reference. That there are two indications it is wrong, that ices it for me.

So, this church has in it those who hold to these false teachings. That is not unique. The church today has many who hold to false teachings. I cover some of them in another book, [14]"Destructive Christian Doctrines – or How the Church Became Irrelevant." What should we do? What should this church have done? I see only three paths for dealing with false teachings.

The first is what God commanded the Israelites to do and what the church in the Dark Ages did, kill them. God commanded it in the Old Testament, fine. The church in the Dark Ages had no such mandate from God and unfortunately, the dark ages church killed many who were true to God because the church was apostate and itself following false doctrine.

Let me be clear, I do not advocate or condone killing to remove false teaching. We are to love those into the kingdom. Those who have bombed abortion clinics for example have violated the teaching of Christ. It is zeal without knowledge.

The second is shun them out of the church. I look at this in the same light. We are to be a place of love.

The third and right action is to remove them from any position where they can harm others with their false doctrine. The first action, imperative and time sensitive is to remove them from any position of formal leadership and teaching. The church should never provide a soap box for any who teach false doctrine. We should never allow the authority or good name of the church to be used to promote that which is not holy.

To carry this out it may require leadership to denounce that false teaching, and even to point out the one in error, but that should be a last resort. If the person can be silenced without being named and pointed out, fine. If not, it may be necessary to make it clear from the pulpit. The person at any juncture may chose to leave, that should never be the goal, but the church leadership must first and foremost, protect the flock. To do less is to be derelict in their duty, but we must never take action that will push someone out of God's

---

[14] Available in print or e-book on Kindle.

house. If they make that choice, we cannot control that. It is a fine line to walk. Many consider it impossible. I will answer that with scripture.

> 1Co_2:16 For who hath known the mind of the Lord, that he may instruct him? But we have the mind of Christ.

If we have His mind, follow in his footsteps those things that would be difficult or even seemingly impossible should be routine. "We have the mind of Christ" is something we should remember, seek him for guidance, and walk in it daily. I will admit, I am not there yet.

Rev 2:16 Repent! But if not I will come to you quickly, and will fight with them by the sword of My mouth.

The message is blunt and to the point, "repent." It is the message to any church or person who is walking out of the will of God. It is the message that has been presented by prophets of old, John the Baptist, Jesus, the early church, and the faithful that have gone before us. We must both repent personally and present that message to the world and the church. It matters not if anyone hears. John the Baptist and those of old prophesied to the ones around them, some heard, some did not. But like Paul, we must be obedient to the heavenly vision.

Rev 2:17 He who has an ear, let him hear what the Spirit says to the churches. To him who overcomes I will give to eat of the hidden manna, and will give to him a white stone, and in the stone a new name written, which no man knows except he who receives it.

I think one of the most important lines in the whole Book of Revelation is, "He who has an ear, let him hear what the Spirit says to the churches." He is saying, "It is important enough that I want you to read this at least three times." He who has an ear, let him hear. Pay attention. It applies to the book of Revelation and it applies to every book of the bible, and every avenue through which God speaks to us, first through the written word, then the word of the pastor, teacher or prophet, just as important, through the still small voice of the Holy Spirit. We must hear, listen and heed His word every time we hear it. We must listen to the voice of God. Too many are looking for the spectacular, not the divine. Look at how

God spoke to Elijah, he got his attention with the loud and noticeable and then spoke quietly.

> 1Ki 19:11 And he said, Go forth, and stand upon the mount before the LORD. And, behold, the LORD passed by, and a great and strong wind rent the mountains, and brake in pieces the rocks before the LORD; but the LORD was not in the wind: and after the wind an earthquake; but the LORD was not in the earthquake:

> 1Ki 19:12 And after the earthquake a fire; but the LORD was not in the fire: and after the fire a still small voice.

> 1Ki 19:13 And it was so, when Elijah heard it, that he wrapped his face in his mantle, and went out, and stood in the entering in of the cave. And, behold, there came a voice unto him, and said, What doest thou here, Elijah?

As a note this is where God told the despondent Elijah who was about to give up that He had 7000 faithful who had not kissed the alter of Baal. Today we need to hear that voice and I believe I say this with the authority of the Holy Spirit, "You are not alone. There are many who have not signed on to the perversion and sin of the world that exists in 2019 and will in 2020. Get up and go in My name." Let us not forget the 'reserves' that back us.

> 2Ch_32:8 With him is an arm of flesh; but with us is the LORD our God to help us, and to fight our battles. And the people rested themselves upon the words of Hezekiah king of Judah.

Back to the voice of God. That voice may come in the form of a still small voice that I have heard at times, reading the word, reading writings by those who have listened to Him and recorded it, and to prophets, pastors, and teachers who have accurately presented the word. I have been privileged and blessed to sit 'at the feet' of those who could do so since childhood. Much has been given me, much is expected. I do not take that responsibility lightly. I look at what Paul said to Timothy and hear it ring to me.

> 2Ti_2:2 And the things that thou hast heard of me among many witnesses, the same commit thou to faithful men, who shall be able to teach others also.

Like Timothy who sat at the feet of Paul and Paul who sat at the feet of Gamaliel, I was privileged to sit at the feet of those who ably presented the word. I am responsible for what was given me.

> Act_22:3 I am verily a man which am a Jew, born in Tarsus, a city in Cilicia, yet brought up in this city at the feet of Gamaliel, and taught according to the perfect manner of the law of the fathers, and was zealous toward God, as ye all are this day.

Paul was taught the law perfectly. As an aside, we see one other picture of Gamaliel in this passage.

> Act_5:34 Then stood there up one in the council, a Pharisee, named Gamaliel, a doctor of the law, had in reputation among all the people, and commanded to put the apostles forth a little space;

Gamaliel was the one when the disciples were being called to task for preaching Christ who said, "Let's wait and see if this comes to naught, least we find ourselves fighting against God." I have seen those over the years that struggled with the gospel, was it true and have seen many who did later come to know Christ. Maybe their path was longer, but they got there. Like Nikodemus, who came to Jesus by night and later helped bury our Lord's body, I wish Luke had done a little more research and told us what happened to these two men, both teachers of the law.

I caution those who are now more mature in age. I am seventy-five years old. I have sat at the feet of one pastor who was about 40 years my junior in years and oh, and by the way, female with red hair. For some that is three strikes against her and I can assure you they have erred in not accepting God's gift. Although she was not the most polished and not in my life for a long time, I gained some great insights from her.

God has been faithful over the years in providing me with those who would feed me with his word. I listed some those who served me faithfully in the dedication of my book on Titus. If you check you might note that some do not have a "Rev." prefix to their name or a bunch of letters after it. One was a school maintenance person, two who I erred and did not mention were a furnace salesman and a press operator in a printing plant. We are all called to be ministers of reconciliation; be that the reconciliation of a man to God or a man to a man. Those who are there, with or without a title, are ones who I felt did that well.

I have had only two short times that I was not being cared for and those were two periods when I was the one who God held in a situation to be salt and light. In both I was in a church where the word was not being accurately presented, it had at one time and the leaders took a left turn, but in one case I was restrained by God to stay for a time, actually, 14 months, it seemed like a decade. The other a period of about six months was when God was telling me it was time to leave and I was telling him, "I don't need another upheaval in my life." Telling God 'no' defines stupidity. He had not defined, move now, but I knew it was soon and I wasn't ready. I was telling him, please don't say, "move now." It was shortly after the loss of my wife and I just wanted some stability. Ironically, the move, when He finally orchestrated it, and He did, created no upheavals and in fact brought peace. I believe it was Brother Oral Roberts[15] who said, "When God does it, he does it right."

Although there is a warning in this passage, there is a message of hope. "To him who overcomes I will give to eat of the hidden manna, and will give to him a white stone, and in the stone a new name written, which no man knows except he who receives it." Those who overcome will be fed, and will be in that number as the song says, "When the saints go marching in." That song has been so perverted by the world, but the message in it is still true. And we overcome by the blood of the Lamb and the word of our testimony. I know, I said that before. Maybe you needed to read it again?

Too many of the churches have taken the songs of battle and victory out of their Hymnals. We need to sing them to God and to one another. Let is tell our fellow Christians. He is Lord Sabaoth, the Lord, mighty in battle. When I researched it, I found two songs that carry this theme and both follow scripture. The first is from the Martin Luther hymn, "Ein Feste Burg ist unser Gott", the English

---

[15] I know there are many who feel Oral Roberts was off base. I have a different feeling on it. I sat in his services, listened to his radio programs and listened to his taped sermons during the 1950's. This was truly a man of God. Most of the criticism of him came from events after the tent services ended. I have looked at this, I am quite sure he walked out of God's will for his life in some of the later items.

translation is better known to us, "A Mighty Fortress is Our God."[16] The third verse proclaims:

> Did we in our own strength confide, our striving would be losing;
> Were not the right Man on our side, the Man of God's own choosing:
> Dost ask who that may be? Christ Jesus, it is He;
> Lord Sabaoth, His Name, from age to age the same,
> And He must win the battle.

I find that wording interesting, "He must win the battle." Let us always remember, the battle is the Lord's. But I see this Lord in battle in more recent music too. A contemporary writer Robert Gay wrote this.

> Lord Sabaoth, the Lord of Hosts
> Mighty in battle
> Lord Sabaoth, Lord of the army of light
> (repeat)
>
> When the enemy encamps round about me
> I will not fear or be terrorized
> For the Captain of the Host is my defender
> And as I praise His name all hell is paralyzed

Again, we see the power and protection of God. We have here two men of God and I have no doubt they both are that, one from today, and one from the 1500's who over nearly 500 years ago proclaimed, God is mighty in battle, He and His army will win. There have been times I was tempted to retreat, to pull back, and there were a few times I did. I am humbled and reminded of what is right, what is prudent, when I read this by Luther.

> I cannot and will not recant anything, for to go against conscience is neither right nor safe. Here I stand, I can do no other, so help me God. Amen.

---

[16] From the Expository Files. http://www.bible.ca/ef/topical-who-is-the-lord-of-sabbaoth.htm

I have asked, "Would Luther recognize and be proud of the church that bears his name today?" I could ask the same of those of other movements. When we are tempted to back up, to run or retreat, to lag behind, to let someone else carry the battle, we should remember this. I liken this to the writing of the Apostle Paul.

> 2Ti_4:7 I have fought a good fight, I have finished my course, I have kept the faith:

> 2Ti 4:8 Henceforth there is laid up for me a crown of righteousness, which the Lord, the righteous judge, shall give me at that day: and not to me only, but unto all them also that love his appearing.

If you don't want to say "Hallelujah" to that you need another dip in the fountain that flows from Immanuel's veins. Take two if you need them. They are free.

I have known many who have gone on before us who I expect to see. Many of them were responsible for laying down the footsteps for me to follow. I pray that the "footsteps that I leave will lead them to believe."

> Oh may all who come behind us find us faithful[17]
> May the fire of our devotion light their way
> May the footprints that we leave
> Lead them to believe
> And the lives we live inspire them to obey

This is a chorus from a Jon Mohr song that says it so well. I should be our commission as we walk in this life. I would be remiss if I did not insert here a verse that tells vividly how to do that.

> We're pilgrims on the journey of the narrow road
> And those who've gone before us line the way
> Cheering on the faithful, encouraging the weary
> Their lives a stirring testament to God's sustaining grace

If we are cheering on the faithful and encouraging the weary, we are doing the work of love He has called us to do. Unfortunately, I have seen the church all too many times attacking the faithful and

---

[17] Find us Faithful by Jon Mohr. Copyright adm. CapitolCMGPublishing.com

discouraging or browbeating the weary. Brethern; this should never be. There is a man at NCCC who does this so well, he is not a pastor, never was, if he read this, he would probably wonder what I am talking about, but he is fantastic at this. Red just sits there in the lobby before church and smiles and greets people. I wish I could do it as well as he does it. I believe that is a gifting from the Holy Spirit and he is not hiding it. There have been times I needed that encouragement. Thank God for him.

I can remember two of the great men of the faith who served here locally being berated for not being spiritual enough. They loved and served God, they worked hard physically and spiritually, in the secular jibs and in the church. They served their brothers and sisters, they just did what would help the work of God. They were not perfect, but they held to the faith. You probably do not know either one, they are Paul Mummert and Ralph R. Brandt, my dad. I only hope that I am looked upon with half the respect that these two men were. I can show a legacy they left that is awesome both in churches they worked to build – physically and spiritually, and in the lives of their children, grandchildren and now, the next generation is coming on. After my dad passed away Paul became someone I could talk to and seek help. He and I taught a class on alternate weeks. I was a willing student on the weeks he taught. As I have gotten older, those who I can look forward to are getting fewer and fewer, but I have found several who God has placed there for me.

And what about the name on the white stone? We talk about something written in stone as permanent. The name will be written in a white stone, a stone of purity, and it is permanent. We are His as beloved children.

THYATIRA

Rev 2:18 And to the angel of the church in Thyatira write: The Son of God, He who has His eyes like a flame of fire and His feet like burnished metal, says these things:

The references to Christ in this discourse are interesting. Based on this verse, the Christ we will see in heaven is far from the pictures we see hanging in churches. I wonder the reaction of the church if a picture of this description, albeit an artist's rendition, were posted there. My son in law, an artist has told me there are some renditions of this by artists who have specialized in this. I looked at them and found them disappointing

The Christ of heaven is not the teacher, prophet, comforter that we see in those pictures. He is not the shepherd, the one who comforted who those who walked with Him saw when He was here on earth as recorded in the gospels.

He is the one who sits at the right hand of the King of Glory. I believe that when we reach that place what we see will not be what we expect based on our thoughts. I do not know what it will be, but I know a couple things. We will not be disappointed, in fact like the song writer said, "I can only imagine." I believe we will stand in awe in his presence, in wonder that he is far greater than our minds can grasp. I saw this awesomeness of God and His creation again from this side of Glory early this year. The prophets proclaimed it, David and Nehemiah, and I got another glimpse of it.

> 2Ch_6:18 But will God in very deed dwell with men on the earth? behold, heaven and the heaven of heavens cannot contain thee; how much less this house which I have built!

> Neh_9:6 Thou, even thou, art LORD alone; thou hast made heaven, the heaven of heavens, with all their host, the earth, and all things that are therein, the seas, and all that is therein, and thou preservest them all; and the host of heaven worshippeth thee.

A space probe, New Horizons that was launched in 2006, flew by Pluto, the farthest known body in our solar system in July 2015 and continued away from earth. Traveling at speeds we have a hard time imagining here, it took about 9 years to get there! On December 31, 2018 it flew by and photographed an even more distant object, Ultima Thule. A TV program was aired on January 3, 2019 detailing the mission. I work with computers, radio and electronics so some things were relevant to me. In the 1960's I built electronics that flew in space, work with computers and networks and I teach radio subjects. I was awed when I learned radio signals from it take 6 hours to get back to earth at 186,000 miles a second and it will take 20 months for the computer to send back the pictures that were taken at the rate of under 1000 bits per second. Most of us have home internets that are 100,000 times that speed. At that speed it would take 518 seconds, less than 10 minutes. Oh, the vastness of this little portion of the universe God created! And this is only a small portion of the universe. At the time this was written this book was about 40,000 words. If it were sent repeatedly from New Horizons, 65 copies would be in the 'ether" before the first letter of the first copy got here.

As an aside, that object, Ultima Thule was not known when New Horizons launched. I was somewhat awed by the technology and the work that was done to locate an object that New Horizons could be maneuvered to pass. It had limited ability to change direction, so the object had to be moving to be close to its current path when they closed. But I can only imagine what it will be like when I see Him. The one who was there at creation must be greater than the creation. I am also mindful of the scripture that says:

> Psa 8:3 When I consider thy heavens, the work of thy fingers, the moon and the stars, which thou hast ordained;
>
> Psa 8:4 What is man, that thou art mindful of him? and the son of man, that thou visitest him?
>
> Psa 8:5 For thou hast made him a little lower than the angels, and hast crowned him with glory and honour.
>
> Psa 8:6 Thou madest him to have dominion over the works of thy hands; thou hast put all things under his feet:

When I look at what has been done, I am mindful that God has created all of this, including man.

I have met and had conversations with Mayors, City Councilmen, and Members of both the PA and US House and Senate. I have met and talked with one of the Tuskegee airmen and several bomber crewmen.[18] I have been honored to be in their presences, even those I disagreed with, and to have had conversations with each of them. But I look forward to one thing. Even if I were to have the opportunity to talk with a president or vice-president, I know two things that top it all. First is the ability I have to daily converse with, both talk to and listen to the King of Kings. The second is when I will be there with the saints of every tribe and tongue when they crown him King of Kings and Lord of Lords. I am sure I cannot

---

[18] World War 2 Fighter pilots.

fathom what will be like, but I know this, it will be worth what it takes to be there.

Rev 2:19 I know your works and love and service and faith and your patience, and your works; and the last to be more than the first.

He tells the church they have done well and their works, love, service, faith and patience have increased. What an accolade.

And note that He sees them growing, their last works being more than the first.

My son's fifth grade teacher, a fantastic teacher and motivator told us he looked for growth. You cannot change where you are now, but you can change where you will be tomorrow. We need to grow some each day, in knowledge and in the word.

Rev 2:20 But I have a few things against you because you allow that woman Jezebel to teach, she saying herself to be a prophetess, and to cause My servants to go astray, and to commit fornication, and to eat idol-sacrifices.

Then we have what I call more than an oops. They have allowed a false prophet to be in leadership, formal or informal, where she can be an influence and cause some to stray, commit fornication and eat idol-sacrifices. They are joining themselves with others in unholy ways and the eating idol-sacrifices puts them into a relationship with the idol. No person can be fully faithful to God and hold to such unholy practices. With this, the devil has a string he can pull and take you off course. Her teaching is bad. What is even more despicable, the church is allowing her to teach this. The church must in a situation like this go to the extent of removing the opportunity to teach and if that fails and he or she continues to teach without the permission of the church, remove her ability to influence the flock. If that means call her out publicly, fine.

There are many who will start the old litany, this is a woman who is teaching, that is what is wrong. Although it is in this passage, I have seen men in leadership pull people off course with false doctrine in the same way. Look at what Paul says to Timothy. Apparently, he too saw it, be it in the spirit or natural, I do not care. Historically in many churches the pastor had to be male, the women were relegated to a less important role, teaching the children. I see two errors in this, the exclusion of women from the pulpit and the attitude that teaching the children was less important. Our pastor does a children's message, in the traditional service there is often only one youngster. I saw something recently and told that young girl, "You are giving the pastor the opportunity to present a message to the church. Keep coming."

> 2Ti_3:6 For of this sort are they which creep into houses, and lead captive silly women laden with sins, led away with divers lusts,

I think it is clear that those who are doing this are men. I will say this here and will not apologize if someone is offended. It is far past time for the church to clean house. It is not a housecleaning with soap, water, brushes and brooms, but one using the Word of God. It is past time for the church to remove that which is unholy, not cover it with white wash and perfume. This must not become a fetish or a witch hunt but a holy housecleaning. It should begin with those things that are not holy being removed and burned if necessary. It should continue with an examination of the liturgy, music and teachings to be sure they are 100% in line with the word of God. It makes no difference if they are historical to the church or denomination, if they are not holy, they must be removed!

Like Moses destroying the golden calf, they must go. And finally, it should be a purge of leadership that cannot totally teach in line with the doctrine of Christ. Their lives should also be such that they do not compromise the message. This does not mean perfect, it however means being open, forthright, working toward the mark. In Titus Paul talks about those on leadership being blameless. This is not perfect, it is bearing no unforgiven sin, both with God and fellow man. I covered this extensively in 'Titus – A Man with a Mission.'

At this someone is going to say, "He isn't perfect." Let me dash that with you. You don't know half of it. You only live with me a few hours a week. I live with me 168 hours a week. Want something tough, try that. I all too well know where I fall short. I have been working on it for 75 years and am not sure I am doing any better now.

> Isa 6:5 Then said I, Woe is me! for I am undone; because I am a man of unclean lips, and I dwell in the midst of a

people of unclean lips: for mine eyes have seen the King, the LORD of hosts.

If you feel like I do, I want to remind you that God calls imperfect people to do His work. He has called me. Let's go to God's response to this.

> Isa 6:6 Then flew one of the seraphims unto me, having a live coal in his hand, which he had taken with the tongs from off the altar:

> Isa 6:7 And he laid it upon my mouth, and said, Lo, this hath touched thy lips; and thine iniquity is taken away, and thy sin purged.

One of the reasons God can use an imperfect one is because His power, in this case through a heavenly being, can take that which is less than perfect and fit it for His use.

I am a Fire Police, I get to wear a coat or vest that says, WMTFD, the initials of the West Manchester Township Fire Department. Yes, I said, I get to wear it. When I reflect on that I am awed and humbled that they would allow me to be a part of that awesome group. They have honored me by allowing me to be a part of them. I want to never bring reproach on them by my words or actions. I belong to a Search and Rescue group and see the same thing there.

This is the attitude I have when I take on the name of Jesus in any form. I do not want to bring reproach on Him or the church. A few years ago, I bought a cross and flame decal for the back window of my car. I placed the order without a lot of thought. For those who do not know, it is the emblem of the United Methodist Church, but it is also, with the cross, is the emblem of my Lord, the flame represents the holy spirit. I didn't know the durability of the decal and the shipping was nearly the cost of one, I bought 5 of them. They laid on the table for over 2 weeks till I got up the courage to put it on my car. The week after I put it on, I took three of them to church, and dropped them on the table in the Sunday School class. I related why I had some concern about putting it on the car.

I am not ashamed of the cross. I was concerned that someone following me in traffic would see it and I would by my driving not represent Christ or bring discredit to Him. Our lives should always reflect him, even when we are behind the wheel. I want to never bring reproach on Him by my words or actions.

Right now, I have increasing concern about the state of the Catholic Church. As I write this the Pennsylvania Grand Jury report was issued, the Cardinal of Washington has stated he will tender his resignation in February when the meeting the Pope has called is convened. He has since been defrocked. This means he cannot hear confession or handle the sacraments.

I will deal with something here. There are practices in the Catholic Church that are not God ordained and need to go. I mentioned earlier about housecleaning in the church. There are several things that are accepted for Catholic Priests that are not of God. Among them are confession to the priest that carries the idea that anything said in the confessional is never to be revealed while the confession of sin to a Protestant Pastor is confidential unless it involves child abuse or some serious crime that is to be committed. This seemingly innocent difference is part of how the priest sex issue grew.

If a priest confessed in the confessional to sexually abusing a child, it died there rather than being used to remove the priest from access to the young and facing legal prosecution. Worse, charges brought against priests were not investigated and punished. In the workplace a teacher who committed these acts would be fired and not allowed to ever work with children. Priests were just moved to other locations.

The celibacy of the priesthood is another link in the chain. It is based on a misinterpretation of Paul's writing.

> 1Co 7:7 For I would that all men were even as I myself. But every man hath his proper gift of God, one after this manner, and another after that.

> 1Co 7:8 I say therefore to the unmarried and widows, It is good for them if they abide even as I.

> 1Co 7:9 But if they cannot contain, let them marry: for it is better to marry than to burn.

What has happened here is exactly what Paul lays out. It is good to marry if they cannot remain single. I believe too many in the priesthood were there, possibly called by God, but could not live the

52

celibate lifestyle. The pressure of not having sex with women, the concept that this would be mortal sin hanging over them, they moved to the alternative. May God forgive those who have driven this ungodly situation down through the years. May he forgive those who have fallen into sin because of the temptation that was too great.

What I see coming is another 'circle the wagons' approach to hide what has not been revealed. Why do I say that? The church has vowed to investigate any wrongdoing. That investigation will be done by people who are a part of the hierarchy, and possibly some who were involved in it! Any good investigator will tell you that group should never investigate itself. With the aspect of the Pope being accused with being knowledgeable of the infractions and doing nothing, an outside 'audit' is needed. If it is not done, this will only grow and do more harm. Even the revelations by the church have often only been done when an exposure by outside investigators was getting close.

Hidden sin is unforgiven sin. Even if that sin in the case of the Pope is only knowledge without prudent action, it is sin. Only if there was no, none, not one shred of knowledge is the Pope innocent. I will only say that I find that very hard to believe. Like Martin Luther who wrote to Pope Leo X I believe this lack of knowledge is impossible.

> For many years now, nothing else has overflowed from Rome into the world—as you are not ignorant—than the laying waste of goods, of bodies, and of souls, and the worst examples of all the worst things. These things are clearer than the light to all men; and the Church of Rome, formerly the most holy of all Churches, has become the most lawless den of thieves, the most shameless of all brothels, the very kingdom of sin, death, and hell; so that not even antichrist, if he were to come, could devise any addition to its wickedness.[19]

I will say one thing for Luther, he didn't sugar coat it. But then, prophets don't. Yes, I called the founder of the Lutheran Church a

---

[19] From Martin Luther (1483–1546). Concerning Christian Liberty. The Harvard Classics. 1909–14. Letter of Martin Luther to Pope Leo X

prophet. I am sure that can be called heresy in some quarters. Prophets present the word as they hear it from God.

If the Pope were to come to me for counsel at this time, and I am sure he will not, I would strongly suggest, "Contact the leadership of ten of the largest Christian and Jewish Spiritual organizations in the world. Ask them to prayerfully and honestly conduct a no holds barred investigation of the church and make the result public with no redactions." But it would require the church leadership recognize other church entities as valid and pride, guilt and fear will not allow that. I believe this is the only way that a full restoration of the Catholic Church is possible. I believe that those who might be viewed as enemies by those inside the church would with humility see the bigger picture, to be honest and forthright, but do it with God's love, trying to restore, not destroy. I believe that with that, no matter how bad it looks, the Church would have as Lincoln said, "a rebirth of freedom."

I believe that without a concrete action in the next year we will see even more reports like the Pennsylvania one, further eroding the credibility.[20] This is playing out now and several months after that line was written, more has been exposed, more has been revealed and more damage done. I believe openness would head off that. I would personally not care if the Catholic Church or any church organization dissolved today however, I do not want to see that because of the collateral damage it would do to churches that are proclaiming the word and the credibility of those. I would much sooner see repentance, humility and restoration.

Least I look partisan I will state the following. If the church involved was the United Methodist Church, or Assembly of God, the ones I have been a part of, rather than the Catholic Church, I would be saying the same thing. I have also said that of the FBI and the Justice Department. There will never be a valid internal investigation of a leadership problem.

One of the best handled problems in church leadership was in a local church I was involved in. They contacted an outsider, Chuck Clayton and the group he led, which has now been merged with

---

[20] We just had another priest hit with charges and attempted to resign. The resignation was rejected by the Pope.

Christian International, a move I am not sure was good, and they sent three men from the outside to look at the problem. I got to know two of them well during that time and had further exposure to one of them later. They were impressive men of God, wise and discerning. They were also no nonsense. I would have expected such a response from Chuck. They interviewed the involved parties, the leadership and as many of the members as they could schedule. They did not refuse to interview anyone who wanted to be a part of it. The issue that brought the query was resolved and two other issues in the church surfaced in the investigation that would have some day caused problems. Concrete actions were recommended and taken. No internal investigation would have had that impact. People were free to share with men they trusted. They did. Things were exposed and once exposed they can be dealt with.

Earlier I used a phrase, "leadership, formal or informal." There are those who lead, who are appointed, elected, selected and have titles. These are the formal leadership. There are also those who by their walk with God, personal charisma, work, or whatever are looked upon by those in the church such that they have the impact of and at times even greater than that of the formal leaders. All leadership can be good or bad. But the informal have the higher risk of being off the wall and do things that can be damaging. Sometimes the less rational you are the more people listen. When these function in a Godly manner, they are awesome support to the church. I have seen some that were far more fruitful than the formal leadership. I will mention here, sometimes the formal leadership position and the title become a straight jacket that prevents the person from blossoming. If this happens to you, seek God on it. He can break the straight jacket, and he can also say it is time to resign! I have seen both. When the informal leadership walk out of the Godly place, they become a terrible drag on the body's fruitfulness.

Good formal leadership will work with the informal. Poor formal leaders will fear the informal and try to suppress them, even when they are operating in a Godly manner. Often these informal leaders are ones who just want to do what needs to be done, don't want the hassle of board meetings, just want to do God's work. They can appear to be threats to formal leadership. Fear drives the situation. I know personally I prefer to not be on boards, to just be able to do what I need to do. I will share elsewhere how that can cause consternation in formal leadership that is insecure.

I am sure I drilled deep on that one.

Rev 2:21 And I gave her time that she might repent of her fornication, and she did not repent.

Here is a fantastic show of God's love. In spite of the false teaching and the damage this one is doing to the church there has been a time to repent. Often when we see something going wrong, we want to get out the axe and wack off heads. I have seen this done and in most cases it has been like the wheat and tares, the tares get pulled up but some of the wheat is lost in the process. Although I am one who sees a problem and wants to get it fixed, then move on to the next problem and fix it, (repeat) I can usually, but unfortunately not always, take a more careful look before 'starting the executions.'

The church needs to remember that it is the church, not an organization, it is the body of Christ and it is incumbent on it to show the love of Christ in all, repeat, all its actions. Showing that love is on one hand, protecting the flock while also not destroying someone, even someone who may be in error. Jesus went out for the one hundredth lamb. He also didn't hesitate when he ran out the money changers.

I will share an unrelated thought here. I have been in churches that were adamant that the church never sold anything in the church. It was because of this act by Jesus and quite frankly, I would prefer the church at the least limit its engaging in commerce. However, it is appropriate at times and I am not dogmatic about it. It should not be an extension of Wal Mart but often it is appropriate. For example, if people buy books for classes, they have as I call it, skin in the game and generally put more into it. On the other hand, we need to be sensitive, at times the one who needs that book or class cannot handle the cost. Allow me to suggest that the Church have a policy of either giving a book or reducing the cost in that case. I personally have gone to a leader of a group and advised that if they had someone in the class that he or she felt would be disadvantaged by the book cost, let me know. I do not want to know who got the free book and just as important I do not want them to know who paid for the book. Why? I want both of us protected. What if a week after I paid for a book for John who was saying he could not afford it I saw him do something that indicated that was not true, valid or not? It could harm relationships. What if John knows I paid for it? He will feel he must treat me differently. I would prefer he not know; hence he must show love to everyone in the body! Yes, I am sometimes sneaky, but I try to use it to promote the body of Christ. I have several times overheard someone telling another person that someone paid something for them and commenting that people here really care. That couple of bucks I gave for that was well spent.

One morning I approached a man in the church I respected. I asked if he could give an envelope to a couple who were struggling. There was money in it. He looked at me kind of quizzically and pulled out an envelope. He responded. "Sure. Could you give this to them."

God had spoken to both of us to do something, by exchanging envelopes we did not reveal the donor, something neither of us wanted to happen. So much good can be done if nobody wants the credit.

I run with a Search and Rescue team. There are several teams in the area with varying resources, varying experience, varying training. Each has value but two teams in particular who have the least experience in leading a search seem to want to run the show. Ego takes over. They would be best to let the most experienced management team on site, generally from two of the mature teams run it, put one or two of their people on the management team as support and to learn, and provide resources elsewhere. I joked after one search where they were trying to lead and were causing confusion that we need signs, "Park your car and ego here." I am not sure these would be inappropriate in church parking lots.

Back to selling in the church. No, I didn't forget about it. I may get sidetracked[21] at times, but I get back on the main line. Jesus said. "My house is a house of prayer, you have made it a den of thieves." He did not say, "You have made it a place of commerce." He said, "You have made it a den of thieves." I have heard some say they were cheating the people in the money changing and I will buy that interpretation as valid however, I further believe they were stealing the Word of God, the best of God, His Power from the people. The

---

[21] Sidetracked is a railroad term, it is putting cars or a train on a 'side track' to allow another train to pass. When it does, the train gets back on the main line and proceeds.

house of prayer had become a den of thieves, not only the money changers, but also the priests. When we do not present the good things of God to the people, we are stealing that from them. In that I believe we have turned the house of God into a den of thieves. If we properly represent Him, that house is a place where he will dwell with us, lives will be made better, people will be blessed.

Rev 2:22 Behold, I am throwing her into a bed, and those who commit adultery with her into great affliction, unless they repent of their deeds.

When I first read this, I missed the last line. It shows that even with the sin and perversion, God is willing to give more time for repentance. There is an old adage, haste makes waste. Too often haste in the church has brought destroyed lives. I know of a church where a hasty act based on pride negatively impacted fourteen people, seven couples who were effective leaders. Over 20 years later, few of them have returned to leadership roles, one did about 10 years ago, got slapped around again and just a year ago finally stepped up again. One died without ever stepping up again. Based on a short accounting, the kingdom probably lost about 150 man years of good leadership and ministry on one hasty act. There was no serious attempt by the leader to reconcile with the group.

Even if this group was wrong and I was close enough to know the counsel they gave that was rejected was good. Unfortunately, they were rejected along with the counsel. If you wish to see a model of this act, look at the story of Solomon's son Rehoboam rejecting his father's advisors and their counsel. He took on counselors who agreed with him. It maps well. He lost ten of the tribes of Israel.

> 1Ki 12:6 And king Rehoboam consulted with the old men, that stood before Solomon his father while he yet lived, and said, How do ye advise that I may answer this people?

> 1Ki 12:7 And they spake unto him, saying, If thou wilt be a servant unto this people this day, and wilt serve them, and answer them, and speak good words to them, then they will be thy servants for ever.

> 1Ki 12:8 But he forsook the counsel of the old men, which they had given him, and consulted with the young men that were grown up with him, and which stood before him:

> 1Ki 12:9 And he said unto them, What counsel give ye that we may answer this people, who have spoken to me, saying, Make the yoke which thy father did put upon us lighter?

> 1Ki 12:10 And the young men that were grown up with him spake unto him, saying, Thus shalt thou speak unto this people that spake unto thee, saying, Thy father made our yoke heavy, but make thou it lighter unto us; thus shalt thou say unto them, My little finger shall be thicker than my father's loins.

> 1Ki 12:11 And now whereas my father did lade you with a heavy yoke, I will add to your yoke: my father hath chastised you with whips, but I will chastise you with scorpions.

One thing I have seen over the years, when those who have over time given wise and prayerful counsel, do not fire them just because they disagree with one of your current pet ideas. You may be the king, the leader, but rejecting wise counsel is foolish. Even if they are wrong, and it is possible, reason with them. It may be both you and they are wrong. That process with prayer may result in a Godly direction that brings success.

God has allowed a time for repentance. I believe that is the usual for a loving God. In the case above, there was ample opportunity for reconciliation, attempts to reconcile from group, promises were made and not kept.

Rev 2:23 And I will kill her children with death. And all the churches will know that I am He who searches the reins and hearts, and I will give to every one of you according to your works.

The first line was at first to me a mystery. Are her children natural offspring or those who are her children in the heresy? The same Greek word that is translated children here is the one that is translated children in 3 John 1:4.

> 3Jn 1:4 I have no greater joy than to hear that my children walk in truth.

It would seem clear here that the children John is speaking of are not his natural offspring but those who he has mentored and brought up in the faith. Although those 'offspring' of Jezebel are being led

59

away from the true faith it is very valid to call them her children. I believe it refers to those who have been brought into her error.

The next line, "all the churches will know that I am He who searches reins and hearts" is using a very old meaning of the word reins. The Greek word refers to 'the inmost mind', Webster defines it as "the seat of the feelings or passions". It is what directs us and drives us.

I'm going to digress here to show an important point. There are many who are in the "KJV only" group and there are others who tout one or another English Translation of the Bible as the best to the extent that some imply that other translations are wrong and even some consider them heresy.

I got a glimpse of the error in dictating versions when I was looking for a 9mm handgun. I have friends who are shooters. I asked one what weapon would be the best. He mentioned that he liked a certain brand and model, but then responded, but the best is the one that works best for you. After asking a couple things he recommended a couple models, none of which were his favorite. He is a big guy and told me that the one he uses may be too heavy for me to be comfortable with. In many areas including the scriptures, the best is what serves you. If the KJV does, fine. If Good News for Modern Man does, use it. Encourage others to read and study the word. My shooter friend told me, it isn't any good to you if you are not comfortable with it and you do not practice. He shared another truth here. To know the bible, you have to 'practice'. When you do the language and style become comfortable.

There are few today who are comfortable and capable reading 1615 Elizabethan English, the language of William Shakespeare and the KJV. For them, and I am one of them, the KJV is great yet I look to other translations because there are words that have changed meanings over the years. I look for places where a word or phrase changed and then go to the lexicon to see why. Most times the words are changed because the connotation of the word and sometimes even the meaning have changed.

The English language is not static and worse, there are multiple dialects. The fuel for a car in the US is gasoline or gas, in England it is petrol. The cover for the engine in the US is a hood, in England a bonnet. To table a motion in the US is to put it aside, in England it is to bring it up! This nearly caused a major rift in the planning for D-day when the Brits wanted to 'table' a motion. The Americans considered the resolution vital to the success and a war nearly broke out till an American came in, saw the issue for what it was and cleared the air. Check the word that is translated 'angels' in the KJV Psalms 8:5.

> Psa 8:5 For thou hast made him a little lower than the angels, and hast crowned him with glory and honour. (KJV)

Then compare it with this.

> Psa 8:5 For You have made him lack a little from God, and have crowned him with glory and honor. (MKJV)

The word that is translated angels in the KJV and God in the MKJV is the Hebrew el-o-heem' which in Genesis 1:1 and in many other references is translated God.

> Gen 1:1 In the beginning God created the heaven and the

earth.

I am far from an expert on the history of the English language, I am sure Miss Hackman my senior high English teacher would roll in her grave with my putting the words expert and English in the same sentence, but I am somewhat cognizant and can use older dictionaries to help me. I get a clearer picture of the scripture when I look at what multiple translators have done. I find three verses in the Psalms that tell me this is valid.

> Pro_11:14 Where no counsel is, the people fall: but in the multitude of counsellors there is safety.

> Pro_15:22 Without counsel purposes are disappointed: but in the multitude of counsellors they are established.

> Pro_24:6 For by wise counsel thou shalt make thy war: and in multitude of counsellors there is safety.

In a multitude of counsel there is safety. Looking at the work of several groups of translators can bring more understanding. I also look to Peter who writes:

> 2Pe_1:20 Knowing this first, that no prophecy of the scripture is of any private interpretation.

When someone comes to me with a 'truth' only he knows, be he the church janitor or the Pope, I will cite this scripture. When someone cites one verse as the basis of a teaching without any support, I am reluctant to embrace it. I want the teaching to be supported by more than one scripture. Does it hold in line with the scripture as a whole? When I see a complex teaching, I am reminded of scripture that shows the way to God is not complicated.

> Isa_35:8 And an highway shall be there, and a way, and it shall be called The way of holiness; the unclean shall not pass over it; but it shall be for those: the wayfaring men, though fools, shall not err therein.

The meaning of fool is not one who is foolish but one who is not as we would say, the brightest. God has made the plan for our lives simple. It would not be consistent with other concepts of God. If it is God's will that all come to repentance, how can that happen unless the gospel can be grasped by all?

To conclude this rabbit trail, which I am sure is needful, the best bible is the one you can understand. No individual has truth no others have and the way of holiness is simple.

The verse concludes with the line that He will give to everyone according to their works. We need to do the work He has called us to, not to be righteous but because we are righteous. We need to do the work He has called us to, not what He has called someone else to do.

When we do what someone else is called to do we push them back and steal their opportunity to serve God and quite frankly we steal the best from that person that would be served. We can never do as good a job as the person who God has designated for a task.

We must remember that God will reward us according to our works. I believe that happens here. I believe it will be in Heaven too.

Rev 2:24 But to you I say, and to the rest in Thyatira, as many as do not have this doctrine, and who have not known the depths of Satan, as they speak, I will put on you no other burden.

But here it is clearly said, there is a remnant that has not fallen to the evil in their midst. They have kept the true faith. This is something I will mention several times because I believe it to be very important.

When Christ comes there will be a remnant that is faithful. We see this in Thyatira and in other churches.

When we have the urge to wring our hands and say all is lost as we see churches heading into deeper and deeper apostasy, we need to remember that there are faithful men and women in this world and God may have some of those seeded, even in the most apostate churches, not necessarily to bring that church out of the apostacy, but to provide a lamppost to those who may be saved out of that situation. They may be the ones who stay while the others they are rescuing are leaving.

When an aircraft is being evacuated, the flight attendants stay till "the fire is too hot, the water too deep or the smoke too thick for them to help others get out." Only when they can save no more can they leave. I believe there are those in apostate churches who will stay, at God's direction, until figuratively the fire is too hot, the water too deep or the smoke too thick for them to effect rescues. And like several stories of flight attendants who have given their lives in that situation, I believe there are those who will not be relieved but will be retired directly to glory from that situation to be honored heroes of the faith.

If there is one thing this study of Titus and the Seven Churches has done, if it does not speak to anyone else, it has resulted in a change in me. I know there are people who know me who will shake their heads and say that is impossible. I now view churches and denominations very differently than a year ago. A local church or denomination may have gone over the hill and become apostate. They may be steeped in Satan's work. But if there may be those who are there who have remained as light posts, they could be the ones who may show men and women the way to meet God. Those newer ones may leave while the others stay to minister. If they provide that, through them God is still using that organization, even in its depravity.

I believe God places a wall of protection around these who are there to fight the battle. You can't tell me He doesn't. I have been there. I know. If he really wants you in that place of rescue, you will have his protection.

Rev 2:25 But that which you have, hold fast until I come.

If there is a message to all of us it is here. "Hold fast until I come." It carries two very important messages to the church, first, "Hold fast." Keep the fire of the devotion to Him burning.[22] Keep the lower lights, the ones that guide ships into the safety of the harbor burning.[23] It is finished with the promise, "Until I come." One more time he reminds us, He is coming back! His return is sure! The songwriter said, "This hope we cherish not in vain, but we comfort one another by his word."[24] The early church greeted each other with the word, "Maranatha', the Lord has come. And with that coming, he left here with the promise of his return. We, more than two centuries later are awaiting his return. I do not know when that will be nor will I speculate but this one thing I know for sure with authority, we are two thousand years closer than Peter, Paul and John were. Another songwriter[25] penned,

> This could be the dawning of a grand and glorious morning
> When the face of Jesus we behold,
> The saints of every tribe and tongue are waiting his returning
> And this could be the dawning of that day.

It could be today, tomorrow, next week, or it could be another thousand years. I have seen so many theories and so called 'prophecies' fail over the years on the Second Coming that I no longer look to them. They are misguided at best. I am close to saying they are of Satan. I look to the scripture. There are some things I know for sure. He is coming back. It will be as a thief in the night, that is with no warning. Mark that one who says he knows and pay him no mind in anything he says. He or she is a false prophet, condemned out of his own mouth. If you wish to see if I am correct, check out the many failed prophecies of E. G. White.

The book "Selling Fear" the writer, a historian says that for generations people have thought they were the one to see the world end. I have studied historical newspapers and we see it back at least to the time of the American Civil War that produced the song that began, "Mine eyes have seen the glory of the coming of the Lord."

---

[22] From "May All Who Come Behind Us Find Us Faithful" John Mohr
[23] "Let the Lower Lights be Burning." Philip P. Bliss
[24] "He's Coming Soon." Words by Thoro Harris, 1916. Music Adapted from Aloha Oe, by Queen Liliuokalani of Hawaii
[25] "This Could be the Dawning" BILL GAITHER

World Wars I and II were thought by those who lived through them to be Armageddon. As a six year old boy I can remember the beginning of the Korean War and hearing people talk of how we were headed into that final battle. Seventy years ago, many in the church were sure that Armageddon was upon us and in fact in progress. With many this was not a maybe, this was for sure. The final battle was in progress. Seventy years and over 100 wars later we are waiting. But today could be the day our savior appears.

I have seen writings that talk about earthquakes being more numerous, but a few years ago I looked at the statistics and the highest amount at that time was in the 1940's. When someone writes something like that, I check. All too often, it doesn't pass muster.

We humans are fixated with time. We fail to see eternity. I look at existence as a time line. In Geometry a line has no defined length, it is infinite, we only see the portion drawn on the paper but that line extends to infinity in both directions. And that concept is difficult for even some very astute students of Mathematics. Likewise, God's time line is infinite. If we draw a time line from Adam to today, about 6000 years, that line extends both ways, in the past to eternity past and in the future to eternity future. Our time on earth here is like the visible line on the paper, only a small portion of the whole is shown.

As a Mathematician I have studied two concepts that hold so much of the attributes of eternity; infinity, and a line. Infinity in math and the line in geometry are ways of describing something that has no end. And I am just as sure that I have not grasped all of it, maybe not even more than a small portion. When I teach on infinity I ask, what is infinity plus 1? The answer, infinity. When I talk of eternity, how long is eternity minus one thousand years? Eternity. Eternity plus 10,000 years? Eternity.

Rev 2:26 And he who overcomes and keeps My works to the end, to him I will give power over the nations.

He who overcomes. He who keeps My word to the end. What a statement. How do we overcome? By the blood of the lamb and the word of our testimony. It is important that we testify that He is Lord and Lord of our lives.

> Rev_12:11 And they overcame him by the blood of the Lamb, and by the word of their testimony; and they loved not their lives unto the death.

We are to keep His works to the end, till He returns or He calls us home.

He will give that one power over the nations? This throws me a curve. At that time, what nations? Maybe, just maybe there is something that is in scripture about beyond the rapture that we do not understand? Is there something that is either not covered in scripture or is it that our finite minds cannot grasp it?

> 1Co_2:9 But as it is written, Eye hath not seen, nor ear heard, neither have entered into the heart of man, the things which God hath prepared for them that love him.

I look at my life, the amazing things that God has done for me and my family. Maybe there is more than I know and can grasp. I try to not put God in a small box, my cranium. He doesn't fit. But the Holy Spirit can reveal things to you plainly, so you understand. This goes beyond the spiritual things, remember that the God we serve is the one who spoke and this universe came into existence. He formed us. From the galaxies and stars to the human cell and DNA, He created it. He created electromagnetic waves and gravity. The Holy Spirit has access to all of this. He is with us, if we desire it.

Those who use drugs to 'take a trip' and 'expand their minds' have no idea how much allowing the Holy Spirit to work will impact a person. Try this. The next time you are in a situation that is bringing confusion, take a couple minutes to invite the Holy Spirit into your thoughts. He will bring peace and comfort. He can also bring clarity and wisdom to the situation. I have seen this is things from interpersonal relations to understanding a complex issue in science or computers. Remember, he is one of the three in one who was there when all of this was created.

Rev 2:27 And he will rule them with a rod of iron, as the vessels of a potter they will be broken to pieces, even as I received from My Father.

Who is the 'He' that is being referred to? It says in the previous verse that he who endures to the end, will be given power over the nations. It must be those in Thyatira who have endured and been faithful. There are many questions I could ask, "What nations?", "Why?" One thing I know. There will be those who overcome who will rule.

I have heard people say, "God said it, I believe it, and that settles it." I am far more simplistic. "God said it." Whether I believe it or not, it is settled. We humans get so self-important. We play at being God. I am reminded of several scriptures but the one that comes quickly to mind is:

> Mat 6:27 Which of you by taking thought can add one cubit unto his stature?

Even so-called great men have come up wanting. Adolf Hitler at one point ruled a large portion of the world and proclaimed his government, the Third Reich to be "das Tausend Jahre Reich"; in English "The Thousand Year Kingdom". His reign lasted 14 years, 986 short of the thousand. I am a student of that portion of history when some of the greatest evils known to man were performed by the Nazi and Japanese governments. I can show at many critical junctures God's finger was on the scales against these. Hitler touched the Jewish people, the apple of God's eye.

> Zec 2:8 For thus saith the LORD of hosts; After the glory hath he sent me unto the nations which spoiled you: for he that toucheth you toucheth the apple of his eye.

> Zec 2:9 For, behold, I will shake mine hand upon them, and they shall be a spoil to their servants: and ye shall know that the LORD of hosts hath sent me.

Tojo and the Japanese threw in with Hitler and supported him. Together they killed over 40 million civilians in their murderous conquest. The number of combatants sacrificed was even greater.

During those years of conflict I have seen the hand of God. At Midway the Japanese were in confusion, at Normandy the German High Command was blind to the Allied intent for some critical hours. When you study some of these in detail and you see what happened you get the images of the battles in the Old Testament where the hand of God touched a situation. You can call it luck, whatever, but there is nothing left to chance if you believe in God. I personally don't buy lottery tickets. I look at them as a tax burden on the poor because all too many of those squander their money in what they incorrectly see as a way out of poverty. Recently there has been a big flap about a 1.5 Billion (with a B) dollar lottery ticket that was sold and has not claimed. The clock is ticking down on the winner claiming the prize.

I have mused a little about this. What if I had bought that ticket? I know I didn't. This may surprise you, but I would have some trepidation about turning it in. I wonder if the ticket is lost, just not checked, or do we have someone who is sitting there holding the ticket and saying, "I don't know if it is worth the hassle." I think God decides who wins the lotteries too. I think I could do well with that money, but I am sure it would seriously complicate my life. I have never bought a ticket, I have gone in with a group I worked with, I called it insurance. I didn't want to be the only one there when everyone else retired.

Rev 2:28 And I will give him the Morning Star.

I have read his at least a dozen times and not seen anything profound in it other than I am sure it will be something that is of significant value.

Rev 2:29 He who has an ear, let him hear what the Spirit says to the churches.

Again, this is repeated. We are to hear the Spirit.

SARDIS

Rev 3:1 And to the angel of the church in Sardis write: He who has the seven Spirits of God and the seven stars says these things. I know your works, that you have a name that you live, and are dead.

To the leaders of the church at Sardis the message is bleak, I know your works, and they are dead. One of the very few Radio and TV preachers I respect is Ben Haden who retired some years ago. He was awesome at presenting the word of God. As he got older and he has turned his ministry that was called "Changed Lives" to Michael Yousef, another solid man of God. This is a fantastic testimony of a man of God who had the wisdom to pass the torch to a successor. I have over the years listened to them and benefitted from their ministries. In one of his sermons about 40 years ago Ben said, "If you are going to a dead church that is preaching a dead sermon to a dead congregation because your business associates go there, you value your business more than you value God." He restated it, "If you are going to a dead church that is preaching a dead sermon to a dead congregation because your friends go there, you value your friends more than you value God." He continued, "If you are going to a dead church that is preaching a dead sermon to a dead congregation because your parent's bones are buried there, you value your parent's bones more than you value God."

I want to be clear. If these or something like it is the reason you are in this church, it is past time to leave. There is one and only one reason to stay in such a church, if God clearly tells you that you are to stay. I discussed that earlier. That is true with any church, you should be there only if God tells you to be there and leave when he tells you it is time. If you are not hearing clearly from God, get with Him more.

Without that direction, you will eventually be like Lot where your righteous soul will be vexed. If you are there at God's direction, he will provide for your soul to be protected, even if it is a church like Sardis. I have been there, I have lived it, I know. I will also tell you that at some point God may and probably will say, "Time to hit the exit." My wife and I within minutes and in separate locations heard from God, "You are not going back." I heard it closing the church door, she was at home. When I walked in, she looked up from the chair and said, "We are not going back, are we?" I had heard it as the door closed. I asked, "Why do you say that?"

She responded, "I just heard it a few minutes before you walked in."

We were in another church the following Sunday. If you leave a church on God's direction, be with a body, any one, till God directs. We had wanted to leave for fourteen months. God said, "Stay", but when the time of repentance for the leader who had gone into error through pride passed, God told us both independently that it was time to leave. I will challenge you, do not desert a post until God says it is time to go. I mentioned airline flight attendants who are taught that they stay to help others out in a crash until the water is too deep, the smoke is too thick, or the fire is too hot. I will assure you that if you remain faithful to God, he will protect you from the water, the fire and the smoke until it is time to use the exit and there will be a clear flashing sign at the time to make the exit.

Many think the pastor must be preaching some terrible doctrine to be apostate. Over the years I have seen pride and dissention be as terrible. They tear at the very fabric of the church. They destroy lives. Tell me, what can be worse? Wicca? Satanism? No. No. Emphatically no. Anything that keeps the church from meeting the needs of the people, that steals God's best should never be there.

As I write this, I could be facing a similar situation. The denomination that my church belongs to is considering the recognition of gay marriage and recognizing homosexuality as a valid life style. I do not have animosity to gays, they are people who Christ died for, however the lifestyle is not one God recognizes as valid. He calls it sin. I do too. I do not apologize for that. When I was apprised that a proposal was being presented to the denomination, I saw some grave concern in the local body and quite frankly I have concerns too. I sent this to several who were expressing that concern. "We must hold our post till God says it is time to leave, until the water is too deep, the smoke is too thick, or the fire is too hot." We stay until we cannot rescue others. I love that analogy because it says it so well.

Jimmy Swaggert, a great preacher who has had some problems preached a sermon in the 1970's, "I have left this pea patch for the last time." The theme is from the following scripture, the idea that the devil has run me off for the last time, I am staying. God has said I win, not Satan. For those who are not sure if he exists, let me settle that for you. HE DOES. The scripture says he does, he does. In addition, I have seen his evil works, I know HE DOES.

> 2Sa 23:11 And next was Shammah the son of Agee the Hararite. And the Philistines were gathered together into a troop, where there was a piece of ground full of lentils. And the people fled from the Philistines.

> 2Sa 23:12 But he stood in the middle of the ground and delivered it, and killed the Philistines. And Jehovah worked a great victory.

The victory is not ours, it is God's. But we get to share in the spoils but only if we are faithful.

> Act_26:19 After this, king Agrippa, I did not disobey the heavenly vision.

At one level or another in our church we will all have to make the decision, to follow the in thing, what society is dictating or follow God. The church leadership will at some point make that decision, till they do it, as I said, pray, wait and see. They may be weighed in the balances and found wanting. If that decision goes against God's way, then each of us will have to prayerfully make that decision. I will someday stand before Him. It will be my actions that dictate how I am judged; not what others do.

> Isa 5:20 Woe to those who call evil good and good evil; who put darkness for light and light for darkness; who put bitter for sweet and sweet for bitter!

> Isa 5:21 Woe to those wise in their own eyes, and bright in their own sight!

Later: The decision was made to follow God, thanks to our African Church brothers and sisters.

Maybe this will bring some clarity to this situation. Peter Marshall, the 1930's Chaplain of the Senate was told by one of the junior senators who went against his party on an issue that he felt was a moral one and the party was wrong. "It feels good to stand with the wind in your face."

I have though about that at times when I have been in a situation where I was the moral compass of the group, when some or all the others were looking for a comfortable heading and I was standing there, figuratively with the wind and maybe some rain or snow in my face. I am a fire police and I stand at times in wind, rain and snow. But I know in both cases, there will be a time to come into the warm, into the safety. Let me end this with a powerful verse from the Psalmist David, a man who had people at times seeking to take his life.

> Psa 118:6 The LORD is on my side; I will not fear: what can man do unto me?

Admittedly I have gone far afield on this one, but I am sure I have not been disobedient to God.

Rev 3:2 Be watchful and strengthen the things which remain, that are ready to die. For I have not found your works being fulfilled before God.

But even in this terrible situation, there are good things that remain and they are enjoined to be watchful and strengthen the good that remains that will soon die without attention. The angel is speaking to those who remain at Sardis. All too often good things die, not of covert action but of neglect. The church as Sardis is like an injured person who is bleeding out and something must be done to stem that bleeding or death will result. He ends this with the statement that the works are not being fulfilled before God. Their works are not acceptable.

When I look at what many churches are doing, I wonder, are their works acceptable to Him? It is important that we look to God. Our church recently ended a clothing ministry. It was difficult to do but it was not really serving either the church members or the community. It was taking energy, time and space that could have been better applied to other endeavors. In this, I am almost a fanatic. I believe in zero based budgeting for business, the church and the government. In it the new budget does not look at the last, it starts with a blank sheet and only what can be justified for this year is spent. I believe every ministry of the church should be evaluated, maybe not every year but certainly on some periodic basis. Although finances should be looked at, "can we reduce the cost of this ministry and still fulfill the mission?" The people cost, be it paid staff or volunteers should also be evaluated. In business it is the bang for the buck. I do not see this as wrong, in fact I see it as being good stewards in the church. The only difference is in the church the bang for the buck is not a profit on the bottom line, it is lives touched and changed.

I am sure some will say, "Every ministry?" I said, "Every ministry." If the Sunday morning sermon or the worship team are not producing results, why have them? Do they need to be revised? Oh, dear God, I have committed heresy, I should be stoned. You don't pull on superman's cape, you can stop having a morning sermon, but you don't question the worship team. Do we really need a better sound system? If the old one is not working, yes. If not, maybe we do, maybe we don't. Would those dollars benefit the kingdom more if expended elsewhere?

Yes, I said, "Every ministry." Likewise, every expenditure should support the bottom line. Can the electric bill be reduced without reducing the ministry? Our church has installed the motion sensor lights in many rarely used areas to reduce electrical use. I do recommend replacing old lighting on an attrition basis to reduce cost, however replacing fluorescent with LED is generally not profitable, unless the fixtures are bad. Dave Ramsey talks about putting a name of every dollar for a family. When you do that you start looking at each expenditure. I am sure the church would be better served if it looked at finances that way.

The church needs a building, there are things that must be. One local church had bought a building from another church and grew out of it. They financially were not ready to purchase a building, so they rented the local high school for Sunday Morning services for about a year and waited. An old school building went up for sale at a great price. They have a great facility that has even room to grow. The old church building that was built in 1954 that they bought from another church that moved into a new building some years before was sold to another active but smaller church. The question that must be asked, "Is the building supporting or impeding the mission of the church?" If there are no empty seats for visitors, will they come back?

In the late 1940's my family was involved in building two churches. Both started as tent meetings in Army Surplus tents, the ones you see in MASH. They could be bought for a couple dollars and included everything you needed except the three center poles. Locally cut tree trunks covered that. One of those churches is now a residence because those who ran it did not follow God. I find little remains from that work.

The other is still, about 70 years later, a church. And out of the ministry of that church has come pastors, teachers and workers who have done the work of the ministry. It spawned into other ministries. How many people have known Christ directly or indirectly because of that ministry? I have no idea but if the number is less than a couple thousand, it would surprise me. There is one thing I know for sure. I learned to serve God and my fellow man in both the services there and the work in making a place for his word to be preached.

I would ask those who were involved in the church that is now a residence. Were your works pleasing to God? I know of some of the history there. I would also ask, "Were the things that tore that church apart Godly?" I compare what I see of this church, looking back about 65 years to the Church at Greenwood, the one that is still there. Although its past is not pristine, it is first, still there, and second, I can see large amounts of good fruit, lives changed, people brought to Christ, healings, and people discipled to do the work of God.

I will mention here that it is far better that church become a residence than an X-rated movie theatre. I have seen that happen in a small town along Route 6 in northern PA. I fear for many of the older churches. There are two, one in York Springs and one in Mt. Holly Springs that are now divided into apartments. I recently saw that some churches in Harrisburg will be consolidated. Hopefully those buildings will see some use that at the least benefits people.

God is bound to support His work. He is not bound to support our egos.

Rev 3:3 Remember then how you have received and heard, and hold fast, and repent. Therefore if you will not watch, I will come upon you as a thief, and you will not know what hour I will come upon you.

The message here is clear. Remember what you have received and heard. Remember the teaching of the gospel that must have once been rich and pure in Sardis. If they are to remember what was before we must conclude that this awful church had a better past. But it is more than just to remember, it is to hold it fast. It is to cling to the gospel truths that they had once learned. And in that there is one more step, repent. Repent for where you have missed the mark. We will all miss the mark at times. I am not sure if I do it daily or hourly. We must recognize that. But when we miss it, we pick up another arrow, we put it to the bow, we hold steady and with our best we take careful aim, in effect repenting for that past item and with a resolve to make this next one a bullseye.

There is another side to this. If we do not watch our hearts and minds, we will be caught up in the error and we will be judged. We must remain true to Him, His word and His commandments.

His coming is sure, it will be unannounced. We must be ready. We must not be like the foolish virgins who had no oil for their lamps. If we keep our light burning brightly before men; we will have that oil always ready. We will be there when the bridegroom comes.

One songwriter said it well.[26]

> No guilt in life, no fear in death—
> This is the pow'r of Christ in me;
> From life's first cry to final breath,
> Jesus commands my destiny.
> No pow'r of hell, no scheme of man,
> Can ever pluck me from His hand;
> Till He returns or calls me home—
> Here in the pow'r of Christ I'll stand."

I love this song, but this verse speaks volumes to me, it is the essence of a victorious Christian Life. There is no fear in death. It alludes several times to scriptures. "Oh, death where is thy sting?" There is nothing that can separate me from the love of Christ. Paul asked, "who can separate us from the love of Christ?" And I will stand in His power until he returns or calls me home.

Rev 3:4 You have a few names even in Sardis who have not defiled their garments. And they will walk with Me in white, for they are worthy.

I do not know if this scripture is for me to now be aware of, but over the last few months God has impressed me with it such that I have shared it in at least a half dozen settings. It may be because I could be facing a decision I mentioned earlier. It may be that I am to make others aware for their benefit or it could possibly be all of these.

This is the scripture that God impressed me with when I saw others thinking about a quick exit and me getting some thoughts about the same. I had some carnal thoughts that I would have to explain how I could stay and support the decision. The question came to mind, "What would others think of me?" That realization was a big ouch.

---

[26] 'In Christ Alone' Song by Stuart Townend

I was looking to my identity in what others think of me, not what God was saying. I was concerned that if certain events unfolded, I would have to also make an exit to avoid that. That scripture tells me, if God says to stay, I can stay, I can remain pure to him, I need not defile my garments to stay if it is His will. I am still not sure of how this can be, but I know for sure this one thing. If God wills it, and He says it, it is possible. What looks to me like a camel going through the eye of a needle is an easy thing for Him. I am sure he can either shrink the camel or expand the eye of the needle.

> Rom 8:38 For I am persuaded, that neither death, nor life, nor angels, nor principalities, nor powers, nor things present, nor things to come,

> Rom 8:39 Nor height, nor depth, nor any other creature, shall be able to separate us from the love of God, which is in Christ Jesus our Lord.

I am persuaded that nothing can separate me from His love. I can be in the storm but unlike the disciples in the boat, the storm need not be in me. With His armor I am protected. If the storm sinks the boat, I am wearing the ultimate life jacket. I must remain in Him, follow the leading of the spirit, for sure, but that is true no matter where I am. We are all responsible for holding that position. I must stand my watch. If I desert that post, who will hold it?

It may help you understand if I relate something that happened about twenty-five years ago. For several weeks every time I started a study the word 'watchman' came up. When something like this happens, it is generally if not always the Holy Spirit preparing you for something. In the natural I tell people, "Take training. You will not now what you need till you need it. When you do you will either have it or it will be too late to get it."

I did a word study on 'watchman', prayed about it but for a time nothing became obvious. During this time Dee and I were installed as elders in a church. We were prayed over and then Melvin Matson, one of the elders took the microphone and spoke. I am sure none of us really understood at that time all of what he said but over the next couple of years it played out. He told Dee that she would become a mother to women in the church, some of them not much older than she. That played out. He told me I would be the watchman in the church, the one who would look at things and see problems where others missed them. A watchman is the one who sees danger and blows the trumpet to warn. In a medieval castle or in the church, the watchman who rouses people from their sleep is not a favorite. Sleeping people do not like to be awakened. And I knew all of this. After the service I jokingly told Melvin that was an evil word. He asked why. Here is a case of a real prophet, speaking the word of God and he does not really understand the word. He is the messenger who presents faithfully the message he was given. This is why I say that prophecies should never be explained. Even Melvin had no idea what he had brought forth. I didn't either, but I had a fleeting glimpse.

I serve as a radio operator in certain settings. I understand that. I am handed a message. I send it as is. When I teach this skill, I tell them, "If a word is spelled wrong, you send it wrong." This is the role of the true prophet, send the message as it is given. The radio operator would say, "Every dot and dash is sent as it should be." Within three years of Melvin's prophecy my role was tested, I passed, some others failed. Through the Holy Spirit I saw the danger, I warned, and the warning was not heeded. The church leadership slept while danger came and wrought destruction. I was not great in this, I was just obedient to the will of God.

But back to the churches. "And they will walk with Me in white, for they are worthy." I have quoted that to at least a dozen other Christians in recent weeks. I get choked up, my voice cracks. I feel the impact of it. "And they will walk with Me in white, for they are worthy." These who have stayed for whatever reason, who have remained faithful to God, who have kept the faith, who have not compromised, they will walk with Him in white, not just because he decrees it but because they are worthy. They are worthy. I would pray that God will say that of me. I pray that He will say that of all who read this.

God leads us along. If we are attentive, He will lead. All too many want the first verse of the song, God Leads Us Along.

> In shady, green pastures, so rich and so sweet
> God leads His dear children along
> Where the water's cool flow bathes the weary one's feet
> God leads His dear children along

Wow. Awesome. The picture of this. He leads us as the Psalmist said, "Beside still waters." But the songwriter saw more of the picture.

> Some through the waters, some through the flood
> Some through the fire, but all through the blood
> Some through great sorrow, but God gives a song
> In the night season and all the day long

I want to point out the ending, "But God gives a song. In the night season and all the day long." We may go through the water, flood and fire, but if we go through the blood of Christ, we will triumph. He leads us along. I commend those who have stood their watch, and encourage you who are coming behind, "stand yours well."

A couple days ago a Pastor, a somewhat nationally known figure who has a church near the Supreme Court in DC, posted on Facebook that we like Bonhoeffer must compromise. I have spent time with him some years ago and at that time had great respect for him. He was talking about two members of the US legislature who he was making glowing statements about. I cringed. One is a strong abortion advocate. The other is nearly as strong and is over the edge on several other ungodly issues. Twenty five years ago he and his brother were active in the pro-life movement. I posted a note that I could no longer consider his advice and counsel good. Another friend who I would have considered a little wishy washy responded, "We can't move the moral goal posts." He took knew and respected this man. He spoke out. I guess he isn't as wishy washy as I thought.

What really got me was him tying Bonhoeffer's evil teaching on compromise to Paul's statement that he would give his salvation for his brothers to be saved, a misappropriation of the scripture. That act is a slippery slope that one starts down and cannot stop. This is not an act of compromise on Paul's part but an act of love. I suspect the atmosphere of DC has corrupted the message this man has been carrying. I pray his mind will be renewed.

We get all tied up in who will make it to heaven. A songwriter said, "Dreams and hopes of all the ages, are awaiting His returning, And this could be the dawning of that day." I like my paraphrase, "The saints of every tribe and tongue are waiting His returning and this could be the dawning of that day."[27]

I believe there will be those there who we would never have thought would be. There is a fictional story about the Pope waiting to get into heaven and as he is in line a man is ushered to the head of the line and processed. When the Pope finally got to St. Peter, he asked who the man was. The response, "He was a cab driver in New York City and he scared the hell out of more people than all of you preachers combined." Unfortunately, I am not sure that with the watered-down sermons preached from many pulpits this is not as far from reality as we may think. The gospel is simple, but it needs to be preached in power and truth.

I hear people bash denominations and I am sure some of them are teaching doctrines that are far from God. But down in the trenches, at the individual level, I have at times found those who simply love God and are standing firm on His word. For whatever reason they are still in that church. I am just as sure that even if it is not for a good reason, if they have made their peace with Christ, I will be seeing them "walk in white."

No, I am not talking the universal salvation, the many ways to God that is being preached from some pulpits and in fact from at least one national TV pulpit. That is error. There is one way, Jesus, if he is the focus of your relationship, you are one with Him.

I will go further. If you are in one of those churches were the gospel is being mangled, get before God. Find out what your role is. It may be to leave. To shake the dust off your feet. It may be to stay. No matter what anyone else says, do what God tells you. But wherever the decision, keep your garments white. Do not be conformed; do not compromise what is good; do not allow man or one who appears to be an angel of light to detract you from the goodness of God.

---

[27] 1971 Gaither Music Company ASCAP/All rights controlled by Gaither Copyright Management.

> Rom_12:2 And be not conformed to this world: but be ye transformed by the renewing of your mind, that ye may prove what is that good, and acceptable, and perfect, will of God.

> 2Co_11:14 And no marvel; for Satan himself is transformed into an angel of light.

Oh yes, Satan exists. Be careful of the one who looks good, smells good, presents as all of this and lacks humility and love. I joked one night that a man who was making a presentation in a School Board meeting was so slick we needed 50 pounds of oil dry to clean up after him. He was so slick he had to be dripping oil. I will be frank, I have seen some like that in the church. Worse, I have seen them pull people away from the true word.

Rev 3:5 The one who overcomes, this one will be clothed in white clothing. And I will not blot out his name out of the Book of Life, but I will confess his name before My Father and before His angels.

The one who overcomes will be clothed in white clothing, pure before God. But how do we overcome? By what means? We must know how if we are to be successful. Look at these scriptures.

> Rev_12:11 And they overcame him by the blood of the Lamb, and by the word of their testimony; and they loved not their lives unto the death.

First and foremost, we triumph through the blood of Jesus. It is His blood that makes us pure before God. We who are unrighteous are imputed his righteousness. I know I have stated it before but there is only one way, through the blood of Jesus. The scripture is clear.

> Joh_10:1 Verily, verily, I say unto you, He that entereth not by the door into the sheepfold, but climbeth up some other way, the same is a thief and a robber

I love the modified one way sign that has Jesus on it. In Romans Paul continues the theme.

> Rom 4:24 But for us also, to whom it shall be imputed, if we believe on him that raised up Jesus our Lord from the dead;

Rom 4:25 Who was delivered for our offences, and was raised again for our justification.

He was raised from the dead for our justification, for our being put into a right standing with God, to be imputed righteous by His righteousness. In ourselves we are not righteous, we are not able to be in a right relationship with God. Because He, Christ is right with God, we are made right with God by believing in Him. And John writes this.

1Jn_2:13 I write unto you, fathers, because ye have known him that is from the beginning. I write unto you, young men, because ye have overcome the wicked one. I write unto you, little children, because ye have known the Father.

1Jn_2:14 I have written unto you, fathers, because ye have known him that is from the beginning. I have written unto you, young men, because ye are strong, and the word of God abideth in you, and ye have overcome the wicked one.

We overcome the wicked one by knowing Him. When we know God, know his nature, we want to emulate that. We may fail at times, but we overcome.

Jesus goes further, he will confess our names before the father. Can you imagine Jesus going to the father and saying, "This is your child, he has confessed his sin, he has walked rightly before you and men?

I am going to step here into a place that I generally avoid, the discussion on whether once one receives salvation they can be lost, i.e. backslide. This statement makes an argument that it can be. "I will not blot out his name out of the Book of Life." If a name can be blotted out, it would have had to be there at one time and then removed. I will stop with one statement, "I never intend to do anything that would make Christ even consider blotting out my name."

Rev 3:6 He who has an ear, let him hear what the Spirit says to the churches.

Listen to the voice of God.

PHILADELPHIA

Rev 3:7 And to the angel of the church in Philadelphia write: He who is holy, He who is true, He who has the key of David, He who opens and no one shuts; and shuts and no one opens, says these things:

He is the one who has made it happen. "I will build my church and the gates of hell will not prevail against it." He will build His church! He has promised that. I will ask each one of you this, "Are we building our church or participating in building His Church?" Let there be no doubt. Hell will assault His church. Men may rise up against her. But if we put our trust in the Lord, none of these will prevail.

I have heard the first portion of a scripture quoted time and again. No weapon that is formed against thee shall prosper." The person quotes it and sits idly by as the devil ravages what us theirs. Read the whole scripture!

> Isa_54:17 No weapon that is formed against thee shall prosper; and every tongue that shall rise against thee in judgment thou shalt condemn. This is the heritage of the servants of the LORD, and their righteousness is of me, saith the LORD.

Complete the thought, read the complete scripture and see your action, your response. "Every tongue that shall rise against thee in judgment thou shalt condemn." We have a part in stopping the gates of hell against us personally and against the church. "Every tongue that shall rise against thee in judgment thou shalt condemn." When those who will be assaulting the church rise up against it, we are to condemn them. I want to be sure some will not take the wrong meaning out of this. To condemn them may be in prayer, not spouting things in public. It could be in public but may not be. In this we need to follow the leading of the Holy Spirit.

The church has for too long been passive. We need to stop wringing our hands like Linus and stand firm in Him. We are not like the beleaguered forces of Britain who were pushed off the continent at Dunkirk, we are the army marching to victory. Our weapons are powerful. We are strong. We can overcome. As the spies said of the land of Cannan, "We are well able to take it."

The songwriter said it well and we have taken this one out of many of our hymnals because it is too military. "On ward Christian Soldiers, marching as to war." And we have taken out of the church the militancy that is needed for us to overcome. I would love to see "There is Power in the Blood"[28] and "Onward Christian Soldiers"[29] put on the worship docket at least once a quarter. As is my usual, I love the third verse of a song. For some reason they have the most meaning. This verse should be sung at least twice. Particularly note the lines, "we are not divided, all one body we." If we are in Christ, if we are really in Him, we are one body.

> Like a mighty army
> moves the church of God;
> Brothers, we are treading
> where the saints have trod;
> We are not divided;
> all one body we,
> One in hope and doctrine,
> one in charity.
>
> Chorus,
> Onward, Christian soldiers,
> marching as to war,
> With the cross of Jesus
> going on before!

In that sign, the cross of Christ we conquer. It is not the physical cross some carry into the service, although I will say that is not a bad idea, to see it as a reminder what He did, what is important is that cross we carry in our hearts and lives. It is the blood that was shed there that washes us whiter than snow.

> Would you be whiter, much whiter than snow?
> There's pow'r in the blood, pow'r in the blood;
> Sin-stains are lost in its life-giving flow;
> There's wonderful pow'r in the blood.

---

[28] "There is Power in the Blood" Lewis E. Jones, 1899
[29] "Onward Christian Soldiers" Written: 1871 Text: Sabine Baring-Gould Melody: "St. Gertrude" by Arthur Sullivan

Would you do service for Jesus your King?
There's pow'r in the blood, pow'r in the blood;
Would you live daily His praises to sing?
There's wonderful pow'r in the blood.

Yes, there are old songs but there are new ones too. Kent Henry wrote this in recent years.

For the Lord is marching on
And His army is ever strong
and His glory shall be seen upon our land

For the captain of the host is Jesus
We're following in His footsteps
No foe can stand against us in the fray

The question I ask, "Are we marching in this army or are we in rest area?" Are we on the battle field or have we surrendered and are now in a prisoner of war camp? Sure, there are times for us to be at rest and refreshment, but we need to be ready for battle when the call comes. The call to battle has been there since the Day of Pentecost.

Put on the gospel armor,
Each piece put on with prayer;
Where duty calls or danger,
Be never wanting there.[30]

We need to put on the armor and not relinquish it. And more important, where we are needed, be there. To accomplish this, we need to take our rightful place as children of God, heirs and joint heirs with Christ.

Rom_8:17 And if children, then heirs; heirs of God, and joint-heirs with Christ; if so be that we suffer with him, that we may be also glorified together.

If the church today would grab hold of the power that has been handed to us, take it seriously, live in the righteousness we have

---

[30] Songwriters: C. Barny Robertson / George Duffield / George J. Webb
Stand Up, Stand Up for Jesus lyrics © Warner/Chappell Music, Inc, Universal Music Publishing Group, Capitol Christian Music Group

been imputed through Christ, and stand firm against every evil, we could change the world. Eleven ordinary men in 33 AD did. Note I said, "against every evil." In this world today we have evils we need to face, abortion that is now infanticide, homosexuality being promoted in the press, films, schools and even churches, all kinds of perversion in the world and worse, in the church. Politicians and business leaders, and worse, church leaders seem to be willing to do anything to remain in power. This is what happened with the Catholic Church, where abuse and sin were allowed to abound, where the sinners and their sins were protected, and the guilty in some cases were allowed to remain where they could create more havoc. Where are we when these are happening? I disagree with Peter. The time is PAST.

> 1Pe_4:17 For the time is come that judgment must begin at the house of God: and if it first begin at us, what shall the end be of them that obey not the gospel of God?

But alas, we are too concerned about committees, programs, even something that is called dynamic worship to be concerned about and learn where our rightful place is before Him and take it. Want to know what I call dynamic worship? It is not the volume, the tempo, the instruments, the special effects, fantastic lyrics, and the other trappings. It is how we lead people to meet God. I was chagrined when one of our sound persons in the church referred to the worship, which is good, as a performance. I was hoping he was referring to it in the generic and technical sense where it is a performance, but not in the spiritual. As an aside, I mentioned it to him since, he looked a little shocked. He said that in a sense, it is a performance but that it has to be worship. I agree, and when it is worship that he helps create, that is the accurate way for this guy to see it.

We look at the accounts in the Old Testament and I sometimes I see us taking them as fables and fiction. They are acts where God delivered His people. It really comes down to one thing, do we believe or not? I love this account from Second Kings. The enemy has sent an overwhelming force and the city is surrounded. They wake one morning to learn they are under siege.

> 2Ki 6:14 Therefore sent he thither horses, and chariots, and a great host: and they came by night, and compassed the city about.

2Ki 6:15 And when the servant of the man of God was risen early, and gone forth, behold, an host compassed the city both with horses and chariots. And his servant said unto him, Alas, my master! how shall we do?

2Ki 6:16 And he answered, Fear not: for they that be with us are more than they that be with them.

2Ki 6:17 And Elisha prayed, and said, LORD, I pray thee, open his eyes, that he may see. And the LORD opened the eyes of the young man; and he saw: and, behold, the mountain was full of horses and chariots of fire round about Elisha.

How often do we say, "Alas, my master! how shall we do?" The songwriter penned.

> Oh what peace we often forfeit of what needless pain we bear
> All because we do not carry everything to God in prayer
> Have we trials and temptations is there trouble anywhere
> If we have don't be discouraged take it to the Lord in prayer [31]

But it is far more than peace we forfeit. By not grabbing hold of our rights we forfeit the power of the Holy Spirit that is ours each day. It is power to stand when others are falling. It is our power to stand to help others from falling. It is easy to succumb when alone.

The Army puts two people in a foxhole together. Why? Increased firepower, yes. Better surveillance, yes. But one of the things that is there, they support each other emotionally. They will stand and fight together. I believe that is part of God's plan for a husband and wife in a family. When they stand together in God's will, everything from hell can be thrown against them and they will stand. In every venue we need each other. Although I have not had the intimate relationship like with my wife since she died, I have those who have become closer who have given support. We need these and those of us who are alone need them even more. The big problem is simply this, we all too often want to keep people at arm's length. Right now, just friends and relatives I know I could go to

---

[31] Songwriters: Charles Crozat Converse / Joseph Scriven What a Friend We Have in Jesus lyrics © Universal Music Publishing Group

with a concern, there are probably 30. I know each has certain areas of expertise.

Rev 3:8 I know your works. Behold, I have given before you an open door, and no one can shut it. For you have a little strength and have kept My Word and have not denied My name.

There are several doors and the writer is not clear on which one. There is the door to heaven and the door of opportunity to lead people to Christ. I believe it is the former based on the second sentence that shows they have kept true to Him, but I would never dispute it. The ambiguity is too great and the theological ramifications are far too minute. Put simply, I can't be sure and both are within the realm of likelihood, they both are within the character of God, I believe it is to us to individually decide. The church has spent far too much energy, created far too many divisions and produced no valuable results in disputes like this. Churches have been destroyed, denominations have been split, friends separated, and factions created by arguing over what is not worth the effort. And I challenge you to look at the results, the fruits of such dissention.

As an example, the Branch Davidians were a break away from the Davidians who were a break away from the Seventh Day Adventists, who were formed by people who broke from other denominations. The very name Branch comes from the fact that they were a break-away sect!

The last sentence defines why these have that open door. They have not denied the name of Christ and have kept His word. I would that all of us were to hear that from Him.

Rev 3:9 Behold, I give out of those of the synagogue of Satan, those saying themselves to be Jews and are not, but lie; behold, I will make them to come and worship before your feet, and to know that I have loved you.

In a sense I personally struggle with one item in this verse, that there will be those who will come and worship at my feet. I may be too brash in this, but I feel that I have kept His word. It is not something I desire that others come to bow at my feet. But I know that if we are faithful, we will rule and reign with Him. It is His promise. I do not care what position I hold. To be the lowest person in His Kingdom would be far more honor than to be the highest in any other place. To be in His Kingdom is to serve.

I know those who have pressed for position, for honor. I have not and in both the church and the secular realm, I have had honor at times. In all cases those places of honor were because of service or to allow me to serve. For any who would seek honor in either realm, let him be the servant of all. I have been appointed to a position in the Ham Radio for Eastern Pennsylvania. I stood in a club last week and talked about how they can grow and serve their communities. Yes, I said, serve. But I also told them that I am there to serve them, to help them, to provide support. As I looked over the group of over 70, I reminded them of their responsibility and mine to serve our communities. I am there to serve. I have that same goal in the church.

Recently I was honored for community service. As I sat there, I looked around the room, at tens of awesome people I would have bestowed that honor upon before me. And I realized something. That honor allows me to do more. Like the prophet, pastor, teacher, evangelist, and apostle in Ephesians 4, this place of honor is more important a place of service. We are honored to serve and the honor gives more credibility.

Rev 3:10 Because you have kept the Word of My patience, I also will keep you from the hour of temptation which will come upon all the habitable world, to try those who dwell upon the earth.

The writer says there is a time of temptation coming on the earth. I personally believe it has come and is here. The world is full of temptation but if we read the scriptures and history, there has always been temptation. There was great temptation in a quest for power for Adolf Hitler, Joseph Stalin and a host of others. I do not know if it will grow, I expect it will. I also know that God will protect those who will commit to follow Him.

> Mat_24:24 For there shall arise false Christs, and false prophets, and shall shew great signs and wonders; insomuch that, if it were possible, they shall deceive the very elect.

Daily men and women are falling into temptation and error. One of the things that is absent from many churches is accountability and even in those that are doing it, few participate. This is not harsh brow beating but a loving relationship where men or women get together to share their joys, successes, sorrows and failures and work to help each other remain faithful to God. I would discourage these sessions being mixed genders. This is something that cannot be done in a large setting, I recommend that when a group gets to 10, you split it. It can be done formally or informally but the best is a regular time because without that it becomes slipshod and does not fulfill the needs. The scripture talks about iron sharpening iron. Our sessions are on Tuesday evening and I miss about one half of them because of other commitments. They are still very valuable.

> Pro_27:17 Iron sharpeneth iron; so a man sharpeneth the countenance of his friend.

As we meet, share and pray for one another we lift the other in spirit and strength. We have the power of the Holy Spirit and agreement. We have the right to push back the forces of Satan.

Some years ago, Dr. James Kennedy was on Larry King Live. I had viewed the pastor as not being strong and Larry hit him with the age-old question, "Why are there so many hypocrites in the church?" He responded with something that drastically raised my opinion of him. I expected him to wimp out with some form of weasel words. I have repeated of that invalid assessment. I will relate his response as well as I can.

> "Larry, I am not sure how many hypocrites there are in the church or where you got your numbers, but I know there are hypocrites and any are too many, however the reason they are there is because discipline in the church is as dead as the proverbial dodo bird!"

I had to right there repent of thinking this man was a wimp. He had faced the situation and responded wisely.

Larry was left without comment.

As Christians we are tempted to do wrong, to commit sin. But there is a more insidious temptation I see today. It is the temptation to "go with the flow", "to not make waves", to "keep peace." Dag Hammershield, the Secretary General of the UN in the 1960's once said, "Peace at any price only raises the price of peace." I saw that when I was raising children. The more appeasement I was willing to make, the more that was required to get the result. When we barter with Satan or his forces in the world to have peace, to try to have it at any price, the price goes up. Ultimately the price will be our capitulation, our selling our souls and integrity to have peace. I am sure that God is no more pleased with our doing that than he was with Esau for selling his birthright.

> Heb_12:16 Lest there be any fornicator, or profane person, as Esau, who for one morsel of meat sold his birthright.

Rev 3:11 Behold, I come quickly. Hold fast to that which you have, so that no one may take your crown.

Here is the reminder to be ready for His appearing. We will not know the day or hour, and many have fallen into error trying to do so. Wouldn't it be wonderful to have just led someone to Christ minutes before he splits the eastern sky? I hear those who talk about being in worship when He comes, but wouldn't it be far better to be about our Father's business, expanding the kingdom. What if just before he splits that sky, we have just lead someone to Christ and we go up together? We will have ten thousand years to praise Him and then were will be tens of thousands of tens of thousands of years more. Wouldn't it be wonderful to know one more soul made it, or many ten?

I have come to understand the songwriter who wrote:

> Lord I would not stand alone,
> When I come before Thy throne,
> Let me bring at least one soul Oh, Lord to Thee,
> Here I give myself away,
> Take me, use me Lord I pray,
> Let me lose myself and find it Lord in Thee.
>
> Let me lose myself and find it Lord in Thee...
> May all self be slain...My friends see only Thee...
> Tho, it cost me grief and pain...
> I will find my life again,
> If I lose myself, I'll find it Lord in Thee. [32]

---

[32] Fifth verse and chorus of Let Me Lose Myself and Find It, Lord, in Thee by Ross H. Minkler

I have never met the writer of that song, Ross Minkler, but I will some time and when I do, I want to say to him, "you blessed my life." I am sure he has blessed many. Oral Roberts used this as his theme song in the early days of his ministry.

> Oh, the joy of sins forgiv'n,
> Oh, the bliss the blood-washed know,
> Oh, the peace akin to Heav'n,
> Where the healing waters flow.[33]

Yes, there is great joy in knowing our sins are forgiven. If you have never led someone to the Lord, you will find there is a thrill in seeing someone meet Christ. I have only a few times been the one to close the deal but there have also been times where I was the one who set up the situation. I was like the guy who holds the ball for the kicker who makes the winning field goal. He doesn't get the credit for the goal, but he sure has a hand in it being right. I wonder if they recognize, without their accurate work, the goal attempt will fail.

I believe there is a crown for those who lead people to Christ, no matter what role they have in the process. Every role is important. Is the guy who preaches the sermon, the one who walks the person through meeting Christ more important than the one who helped sed up the physical venue, driving the stakes to set the tent, preparing the hall or sanctuary, is he more important? I challenge you, do what you can. As a 4 year old boy I carried tent stakes around to help set a tent. There are two aspects of that, I helped, yes, but I learned to serve. We need to infuse our children with this concept, to serve. There is no greater honor than to be allowed to serve.

I think it is well summed up in the following two scriptures.

> Pro 11:30 The fruit of the righteous is a tree of life; and he that winneth souls is wise.

> 1Co 3:5 Who then is Paul, and who is Apollos, but ministers by whom ye believed, even as the Lord gave to every man?

---

[33] Words by H. H. Heimar, in Tears and Triumphs No. 3 (Louisville, Kentucky: Pickett Publishing, 1902). Music by Leander L. Pickett

1Co 3:6 I have planted, Apollos watered; but God gave the increase.

1Co 3:7 So then neither is he that planteth any thing, neither he that watereth; but God that giveth the increase.

1Co 3:8 Now he that planteth and he that watereth are one: and every man shall receive his own reward according to his own labour.

1Co 3:9 For we are labourers together with God: ye are God's husbandry, ye are God's building.

We need to hold fast to the faith and make Christ known to the nations.

Rev 3:12 Him who overcomes I will make him a pillar in the temple of My God, and he will go out no more. And I will write upon him the name of My God, and the name of the city of My God, the New Jerusalem, which comes down out of Heaven from My God, and My new name.

Here we see the lot of the one who is faithful. He will become a pillar in the temple. He will support the work here. And there is a reward for that one, he becomes branded as one of God's.

I have several insignias I wear to identify the group I belong to. I wear each of them proudly and humbly. I am proud to be a part of groups that help people, a search team, a fire police team, an Emergency Operations Team, New Creation Community Church, etc. There are two aspects of this, the pride in being a part of the group and humility, they allow me to be with them. I mentioned the UMC car decal earlier. I never wish to bring discredit to any of these.

Rev 3:13 He who has an ear, let him hear what the Spirit says to the churches.

Listen!

LAODICEA

Rev 3:14 And to the angel of the church of the Laodicea write: The Amen, the faithful and true Witness, the Head of the creation of God, says these things:

Here he again identifies as the one true witness. Note also that he is being called the head of the creation of God. Remember, Jesus was there, part of the Trinity that was there when God created it all. I find something interesting in the following statement.

> Gen 1:1 In the beginning God created the heaven and the earth.

When I look at the vastness of the heavens, compared to the size of the earth, to put them on a somewhat equal footing in this sentence is interesting. Somehow the earth had to have some special significance to God. It did. It was the place His people would dwell. People speculate if there is other intelligent life. I am a scientist. I have heard the arguments on both sides. I will answer it this way. If God wanted to have life on other worlds, it is there. If not, it isn't. With the vastness of this, the time for even light to transit the distances, I doubt we will know until we see Him.

Rev 3:15 I know your works, that you are neither cold nor hot. I would that you were cold or hot.

Rev 3:16 So because you are lukewarm, and neither cold nor hot, I will vomit you out of My mouth.

I will deal with these two verses together because they are really one thought. I want to remind the reader that all scripture is profitable for our life with Christ. But I do not believe there is a scripture that I have heard preached and commented upon more than this one. I wish I were saying that about John 3:16 or maybe Romans 10:9. If all scripture is profitable, why is this one so heartily embraced? Is it more important? Does it better define the gospel? Or does it meet the criterion of being a preacher's pet idea?

I will show it this way. When I encounter an Adventist, one who is fixated on the second coming of Christ, who talks nothing but Daniel and Revelation I often ask, "If you encountered a person who did not know Christ and you knew you had less than five minutes to convey the gospel to them and you knew they would never have an opportunity to hear the word again, would you spend that precious time discussing the mark of the beast, the feet of clay, the pre, post tribulation arguments or would you go to John 3:16, Roman 10:9-10 and tell him God loves him and point him to Christ?

I can tell you that every time I have gotten one of two reactions, "but they must know about the tribulation, mark of the beast, etc. or "but you do not understand how important it is for them to know about the end time." I have responded, "You have a personal agenda that you hold above everything else. You do not understand how important a soul is." We need to put aside our personal agendas. I get nothing but hostility.

The scripture says it is the goodness of God that leads men to repentance. I believe preachers need to at least once a year preach a real hell fire and brimstone message and if I could figure out a safe way to put the smell of brimstone into the air conditioner at the end of the message, they would get the smell at the end of the sermon, I can't, but I would.

I remember the impact of smell in a three-day IBM conference I attended near Philadelphia in the early 1980's. The conference was one that had some very heavy technical subjects including the announcement of a new IBM data transmission method that was a great advance. I remember sitting there and saying, "This makes one world government possible." It allowed a quantum jump in use of transmission lines. There was one more checkmark on the list of things that must happen before He splits the Eastern Sky. In case you are thinking, "He is an Adventist", forget it. Like all humans I can't see the full list, don't know all that is on it, but one more checkmark is part of the countdown. I don't know how many more checkmarks are needed. I know we are counting down, I hear the ticks of the clock, I just don't know what number we are on. It could be one or one million.

But back to the smell. At the end of the third day of the conference they had a slide show of various shots of the conference (remember this was 1983 and there were no digital cameras, so these had to be film slides that were processed and ready for viewing) with the background music, "you got to stop and smell the roses." The one evening speaker was Isaac Asimov, a forward thinker who studied history and made some interesting predictions that in fact were borne out. His theme was, to predict the future, look at history.

The third morning speakers dealt with balancing careers and family life. When the music stopped and the hotel staff opened the door to the lobby people started out. The lobby had been sprayed with at least a dozen cans of rose fragrance. I actually saw that being done, I had to leave the room during the music and saw about a half dozen of the hotel staff with two cans each, emptying them. The song, you have to stop and smell the roses always brings me back to that, the smell of that room. I would that we could somehow get the church to understand the meaning of hell. Maybe spraying "brimstone scent" in the lobby at the end of the service would help? It saddens me to say that I am sure a significant portion of the church do not believe in hell, Satan and demons. Worse, we are not as interested in being about our Father's business from the time we know to the grave. Look at Jesus, our example. It twelve, in the temple, certainly bending some paradigms, He was doing it.

> Luk_2:49 And he said unto them, How is it that ye sought me? wist ye not that I must be about my Father's business?

I like the TV shows, Major Crimes and the predecessor, The Closer. In particular; I like G. W. Bailey as Detective Lieutenant Louie Provenza, second-in-command of the Major Crimes unit. He is crotchety and cynical. In the story he is not retiring because his ex-wife will get a portion of his pension when he does. As long as he continues to work, she does not get her share of the pension.

In one episode of the series they are encountering a terrible crime scene where their Captain and others have been killed in a shootout in a courtroom because someone loaded a weapon that was brought in as evidence. The defendant was asked to demonstrate something, he took the weapon and started firing. One of the squad comments something about how God could allow this, a question that is often asked in that kind of situation. Louie responds. "I have never had much thought that there was a God, but right now I could be convinced there is a devil." Louie is ahead of many Christians. The scripture clearly talks about Satan and his demons, Jesus even cast them out. But many Christians want to live in la-la land ignoring the scripture.

Allow me to give you a lesson in warfare. I have studied battles from Normandy to Little Big Horn and a lot between. If you do not know your enemy, cannot recognize him, do not know his capabilities, and do not know his location, i.e. that he is on the battlefield, you will lose the battle. That happened to Rommel at Normandy and Custer at the Little Big Horn. That is the mode that many Christians are in, they do not recognize the devil, his capabilities and his location, his allies and far worse some do not believe he exists! They are destined to fail. It would be like a warship denying that an enemy ship is there, even after the shells start to hit! The devil is already firing his guns!

I got a little off track but back to the two verses. God is trying to tell us to remain true and faithful, not just lip service, not just getting by, but by being totally committed.

Oral Roberts told a story about a truck driver who was interviewing for a job. He listened to how each driver ahead of him would tell the boss how he could drive so fast, get close to the edge of the road and not wreck the truck. When he was called, he told the boss, "You do not want me. I can't drive like that." The boss asked how he drives. He responded, "I try to get there as soon as I can, but I don't like to run too fast, I try to stay as far away from the edge of the road as possible. If I am not careful, I might make a mistake and wreck the truck." The boss reached out, shook his hand and said, "You are hired. I want someone who will get my load and truck there and back safe."

It is not how close to the world you get, how fast and loose you can be with the gospel, it his how far you can stay from the world and keep your heart right with God. I believe that is in essence what this says.

These verses became a staple for sermons in the Pentecostal and Holiness movements. They became an emphasis to the extent that they were taking people back to the law with its righteousness by works. The thrust of these sermons was simple, "no matter what you were doing, it wasn't good enough." It was used as what I call, the weekly beating the sheep. After a bad series of meetings by my one boss where it was constant beatings, one of my fellow workers said, "The beatings will continue till morale improves." We need to preach holiness, we need to talk about right living, but it should be in the form of encouragement, not this. I will show the fallacy in this with one scripture passage from Galatians.

> Gal 3:1 O foolish Galatians, who hath bewitched you, that ye should not obey the truth, before whose eyes Jesus Christ hath been evidently set forth, crucified among you?

Gal 3:2 This only would I learn of you, Received ye the Spirit by the works of the law, or by the hearing of faith?

Gal 3:3 Are ye so foolish? having begun in the Spirit, are ye now made perfect by the flesh?

Gal 3:4 Have ye suffered so many things in vain? if it be yet in vain.

Gal 3:5 He therefore that ministereth to you the Spirit, and worketh miracles among you, doeth he it by the works of the law, or by the hearing of faith?

The Galatian church was finding its way back to the law, to works of man, not the work of Christ on the cross. Paul called them foolish. We are not justified by works. We are righteous through his work on the Cross. I know I have said it before, but I cannot say that too many times. It seems that we want to go back to the old ways. If we want to go back, let's do it the whole way and re-institute animal sacrifice!

If Satan can get us to a mentality of works, to the law, he can negate the work of Christ in our lives. We sing it, "Twas Grace that taught my heart to fear, and grace my fears relieved."[34] We leave the church, go on our way and try to be justified by works.

His grace is amazing, it is so vast and so strong to be able to cover the sin in our lives. Sure, we need to walk daily with Him.

John Newton who captained a ship that brought slaves to the US penned that song after he found Christ. He looked at his life, what he had done, how he had failed God and his fellow man, and he saw that Christ had forgiven his sin. I share one thing with John Newton, the awe at the Amazing Grace of Christ. For over 200 years this song has encouraged many.

Rev 3:17 Because you say, I am rich and increased with goods and have need of nothing, and do not know that you are wretched and miserable and poor and blind and naked,

There are many today in the church who do not see the need for God. They have what they need. I will be frank. Without Him a person really has nothing. Many think this is today's society in

---

[34] Amazing Grace, John Newton.

America. That is the message that is preached. And there is some truth in it. Let me show you that this has existed since at least the time Jesus walked this earth. I can also show this existed in the middle ages when the Catholic Church took money to build the large cathedrals in Europe under the false pretense that the donor was buying an 'indulgence', the ticket to allow them to sin. It was an unholy alliance between the laity and the clergy, the laity wanted to sin and the clergy was comfortable with telling them they could buy that right! It was based on a misappropriated statement by Jesus.

> Joh 20:23 Whose soever sins ye remit, they are remitted unto them; and whose soever sins ye retain, they are retained.

The Catholic Church played on this till people started thinking that the they could forgive sins and worse, say a sin wasn't a sin. If you think that went out with the middle ages and the Protestant Reformation, consider the recent initiatives in various groups to say that the practice of homosexuality is not a sin. The bible is clear. Sexual intimacy is only valid with two people who are married, a man and a woman. God said it was sin. No church has the right to say it isn't.

I discussed the indulgences earlier with Martin Luther's letter to the Pope.

You only have to look at the story of the rich young ruler. What did Jesus say to him? Sell, give to the poor. That message as also been perverted.

> Mar 10:21 Then Jesus beholding him loved him, and said unto him, One thing thou lackest: go thy way, sell whatsoever thou hast, and give to the poor, and thou shalt have treasure in heaven: and come, take up the cross, and follow me.

> Mar 10:22 And he was sad at that saying, and went away grieved: for he had great possessions.

But I look also to verse 22. He left with sorrow because he had great possessions.

I watch and listen to few TV and radio preachers. Right now, the only one I spend much time with is Michael Yousef. On more than one Sunday I have heard a sermon from him and then listened to my pastor preach on the same subject. Do you think it might be possible that God wanted me to hear that? Ben Haden, Billy Graham, Oral Roberts (old messages) and K. Fred Price pretty well round out the group.

One Sunday Price was preaching on the rich young ruler. When he read the sentence, "for he had great possessions" he stopped and looked around as he did when he wanted everyone to be paying attention. It is interesting how silence can be loud. There is nothing like a long pause in a sermon to get attention. After the pause he said, "That is where the bible is wrong." Believe me for a couple seconds I nearly flipped out until I realized he was making a point. I knew his commitment to the scripture and for a moment could not believe he said that. After another pause to let that sink in; he said, "he didn't have great possessions", another pause, "the possessions had him." He continued. "Some of you claim you have a problem with alcohol. You don't. The booze has you."

Yes, the young man did have great possessions, but more so they had him. Even to have God, he could not give them up. I have heard preachers use this to show it is Godly to be poor. Allow me to refute that and put a slightly different spin on Zacchaeus. I like to put the scripture here so you can read it, check what I am saying.

> Luk 19:2 And, behold, there was a man named Zacchaeus, which was the chief among the publicans, and he was rich.
>
> Luk 19:3 And he sought to see Jesus who he was; and could not for the press, because he was little of stature.
>
> Luk 19:4 And he ran before, and climbed up into a sycomore tree to see him: for he was to pass that way.

I have heard many sermons detail that he was vertically challenged. I submit the only part that played was what was needed to see Jesus even though he was the chief among the publicans, something I have never heard a preacher mention.

Few know that Fanny Crosby who wrote many great songs, actually poems that others set to music, including 'Pass Me Not Oh Gentle Savior' was blind. But in looking at her songs, and many of her poems were never set to music, I know for certain, in spite of her blindness, in some way she saw Jesus. She lived past 90 years and wrote thousands of poems. Many hymnals have more than a few of her songs and only a relative few were set to music. When I look at the great songs that came from her, I wonder what treasures are in the other ones.

> Luk 19:5 And when Jesus came to the place, he looked up, and saw him, and said unto him, Zacchaeus, make haste, and come down; for today I must abide at thy house.

Allow me to refute the significance of the tree. Do you think Jesus saw him because he was in the tree? The significance of the tree was the effort he made to see Jesus. I will also comment that there has to be some humility for this man of position to climb the tree.

> Luk 19:6 And he made haste, and came down, and received him joyfully.

I believe Zacchaeus had no expectation of an encounter with Jesus when he climbed the tree. He wanted to see Him. But if we reach out to God, we will find that He is already reaching to us. It was what was in his heart that brought salvation.

> Luk 19:7 And when they saw it, they all murmured, saying, That he was gone to be guest with a man that is a sinner.

Hmmm. Like the others standing there were not sinners? I know, they were saying they were righteous and he was in sin. After all, he was a tax collector. He had to be a big sinner.

Too often when a sinner comes to Christ the church takes this attitude. Let me tell you what we should be doing. The scripture says there is joy in Heaven when one sinner comes to Christ. I believe the church should throw a party. Can you think of a better reason to have one? Have a "Sinner came to Jesus Party?" Sounds like an awesome idea. I know, someone is saying, "Yep, he is weird." I confess, I am.

> Luk_15:10 Likewise, I say unto you, there is joy in the presence of the angels of God over one sinner that repenteth.

> Luk 19:8 And Zacchaeus stood, and said unto the Lord; Behold, Lord, the half of my goods I give to the poor; and if I have taken any thing from any man by false accusation, I restore him fourfold.

True repentance results in action. He makes a statement that is misunderstood by many. First, Jesus does not ask him to sell all and follow him, but Zacchaeus spontaneously says he is going to give some away. Why didn't Jesus tell him to give it all? He did that with the rich young ruler. I will submit that Zacchaeus had wealth, but unlike the rich young ruler, at this point, to use the Price statement, the wealth no longer had him. He realized half of it could have more value – to others and him - if given away. The second half of this is also missed by many. They fixate on the "restore fourfold" and miss what is in the first part of that sentence. "If I have taken by false accusation" is a key.

Jesus accepted that. I see something here. By doing so Jesus showed that the collection of taxes was not the problem. It was if he took something wrongfully. He said he would restore it. I would bet that Zacchaeus was one of the better and more honest tax collectors. The fact that he felt he could restore what he had taken wrongfully indicates he was reasonably sure that the amounts were not substantial. It is possible, actually likely he was in fact honest in his dealings and restoring four fold to someone would have been dealing with mistakes, not willful taking. I know, that blows big holes in some sermons you have heard.

> Luk 19:9 And Jesus said unto him, This day is salvation come to this house, forsomuch as he also is a son of Abraham.
>
> Luk 19:10 For the Son of man is come to seek and to save that which was lost.

Jesus then affirms to him and the group that his sin has been forgiven.

When I look at this, I am reminded of one of the greatest verses in any song that exists, the third verse of "It is well with my soul" which is mentioned elsewhere. "My sin not in part but the whole, it is nailed to the cross and I bear it no more." This is how he parted with Jesus. His sin, not in part but the whole was nailed to the cross. He left not carrying his sin, it was dropped at the feet of Jesus. You ask how could his sin be nailed to the cross? Christ had not died. We must remember that God is eternal and timeless. We must get out of the box of time when we deal with God.

When he climbed the tree Zacchaeus was poor, blind and naked. But with this encounter he left a different man. He may have had less than his previous worldly wealth, but the wealth he had gained was far greater.

Allow me to deal with another item, God is timeless. The Holy Spirit was there at creation, "The Spirit of God." Jesus was there in the furnace with the three, when the king said there were four there and one is like the Son of God. The Holy Spirit is the other comforter Jesus mentioned in John 14. He was poured out on all flesh on the day of Pentecost. I find nothing in scripture that says God has issued a recall notice for Him. He is here, he is present, he is available to enrich our lives. I believe the church has replaced the power of God with form and ceremony, and it goes back to where the church lost the power of God. There had to be something to replace it. Like Nadab and Abihu we have substituted the fire of God with false fire.

> Lev 10:1 And Nadab and Abihu, the sons of Aaron, took either of them his censer, and put fire therein, and put incense thereon, and offered strange fire before the LORD, which he commanded them not.

The Holy Spirit is still here. How do I know? First, God sent it, Jesus proclaimed it, and it has not been recalled. Second, I have experienced that power, that strength from Him. I have heard His voice. I know it. I know the value of praying in the spirit. I know he is with me and will be till the end of the age. I can no more deny the Holy Spirit in my life than I can the work of Christ on the Cross.

Rev 3:18 I counsel you to buy from Me gold purified by fire, so that you may be rich; and white clothing, so that you may be clothed, and so that the shame of your nakedness does not appear. And anoint your eyes with eye salve, so that you may see.

Three things are covered here. Gold refined by fire, white clothing and eye salve. Gold is refined by fire to be made pure. When it is heated it melts, the impurities are no longer there. White clothing is the symbol of Holiness. And the eye slave is so they may see the gospel in its fullness.

Look at where this comes from. "Buy from Me." How do we buy something that is priceless. By giving what we have to him. Stuart Hamblen wrote, "Yet the King of all Kings seeks the love of each man."[35] In one line, he defines our relationship with God.

Rev 3:19 As many as I love, I rebuke and chasten; therefore be zealous and repent.

The songwriter Mary Bridges Canedy Slade wrote a song, 'Who at My Door is Standing'. He is standing out there, knocking, will we open? Will we let Him in? A verse ends with the lines:[36]

> "Though He rebuke and chasten,
> He shall with me abide."

It is our choice to have Him abide with us. He will be there if we will just open the door and invite Him to stay. There may be rebuke, there may be correction, maybe even some chastening, but it is for our good. He will be there with us. I have learned this even more in the 12 years since my wife passed away. He will with me abide. My life took a drastic turn. But He was never far away.

At times there has been rebuke. But there is always the reminder that I am His and He is mine. Many do not seem to grasp that truth and wander through life with no compass and no support. I use maps and GPS to navigate. But in my life I have found no better compass than His written word and the leading of the Holy Spirit.

Rev 3:20 Behold, I stand at the door and knock. If anyone hears My voice and opens the door, I will come in to him and will dine with him and he with Me.

Here is the invitation, the beaconing of Jesus to all who hear. He is not physically here. It is our task to present the message to each and every one. But to be there we must be the first partaker, the one who hears and opens the door to our lives.

Rev 3:21 To him who overcomes I will grant to sit with Me in My throne, even as I also overcame and have sat down with My Father in His throne.

---

[35] King of All Kings - Stuart Hamblen
[36] "Who at my Door is Standing" Mary Bridges Canedy Slade

And then he talks to those who overcome, those who turn to Him and love Him. I do not know how many can sit with Him on his throne, and unlike the theologians of old who argued over how many angels could occupy the head of a pin, I will not spend time even thinking about it. But I know, if He says it, it is possible, and it will come to pass. The throne will have enough space.

We are here discussing an omnipotent, omniscient and omnipresent God and trying to describe Him with human knowledge. He is all powerful, all knowing and always present. When we start looking at him that way, our minds are expanded to know Him.

Rev 3:22 He who has an ear, let him hear what the Spirit says to the churches.

God has spoken to His church many times and in many ways. He speaks through His apostles, prophets, pastors, teachers, and evangelists. He has placed them in this world to speak his word to us and lead us to Him and let me assure you, we are not without guidance, they are still here today if we chose to listen. We also have the Holy Spirit, the other comforter that Jesus sent. I know I have stressed the Holy Spirit and mentioned it often, but it is so important in the life of a Christian to recognize this gift from God. All too often the church gets off track, often by those who present a message the congregation wants to hear, not what God has for them.

In the work world I had to set goals. They had to be measurable, realistic and attainable. They also had to support the goals of the company.

I believe the messages in the church should challenge us and lead us to be more faithful to Him. They should set goals for us, measurable, realistic and attainable. I have seen the messages go to the extreme and become what I have called, regular beating the sheep. There is no place for this. In addition, the pablum that does not challenge us to climb more steps is not fruitful and just as important, the sermons that encourage and comfort. I know a pastor has a tough job, but I am also just as sure that if he or she puts the good of the congregation in the hands of the Holy Spirit, allows Him to guide, the stress levels fall. The scripture says:

> Mat 11:28 Come unto me, all ye that labour and are heavy laden, and I will give you rest.

104

> Mat 11:29 Take my yoke upon you, and learn of me; for I am meek and lowly in heart: and ye shall find rest unto your souls.

> Mat 11:30 For my yoke is easy, and my burden is light.

I think too many of the church, including leaders see this as a message to the sinner. It is, but it is to all. We all labor and all are heavy laden at times. He has promised rest. We are to work for Him, the fields are ripe and the laborers are few. Some of us may have to at times do double shifts. But we have the Holy Spirit with us.

Individually, as groups in the local body, and as the local body we must set goals. The five gifts mentioned above can help set those goals by what is presented to the congregation. The pastor has a responsibility for setting the goals for the local body, but they must be done prayerfully, and with input from Godly and prayerful concurrence of the leaders of the congregation. One of the worst splits in the church is when the leaders and the congregation split, what I call a horizontal split. I have included a writing on this as a separate chapter. It was written and posted in a blog and I felt to include it.

Allow me to conclude this. I find it interesting that Paul in his letter to the Ephesians makes this statement.

> Eph 6:12 For we do not wrestle against flesh and blood, but against principalities, against powers, against the world's rulers, of the darkness of this age, against spiritual wickedness in high places.

I have used this scripture elsewhere in this study to show the office of the Apostle is one of service. Too many times when the church encounters problems we are inclined to attack in the flesh. This is the time to take on the devil and his minions and then clean house by repentance and fill the place with the Love that only God can foster.

And so we conclude the study of the Seven Churches.

SILENCING THE PROPHETS

I have looked at this scripture many times.

1Ki 19:10 And he said, I have been very zealous for Jehovah the God of Hosts. For the sons of Israel have forsaken Your covenant, thrown down Your altars, and have slain Your prophets with the sword. And I, I alone, am left. And they seek to take my life away.

When the Prophet of God appears in the midst of his people there is always the risk that there will be an effort to silence him. It has been that way since the days of the prophets of old and it has not changed under grace.

Allow me to explain what the word prophet means because too many in the church today have varying ideas of what a prophet is and most are not even close to the truth. Let me first tell you what a prophet is not. A prophet is not a Christian Fortune Teller. He is not one who will be out predicting the date and time of the second coming of Christ. He will not be one proclaiming himself as a prophet. He will not be proclaimed as a prophet by the mutual admiration society he lives travels in.

If they are not that, what is a prophet? He is one who proclaims God's word to the people. That is it, no more and no less. You may not like what he is telling but if you are allowing the Holy Spirit to work in your life, you will recognize the prophet. And there have been prophets since the early days of Israel and there will be prophets until Jesus splits the Eastern Sky. How do I know that? Look at the word of the Prophet Joel.

Joe 2:28 And it shall be afterward, I will pour out My Spirit on all flesh. And your sons and your daughters shall prophesy; your old men shall dream dreams; your young men shall see visions.

And if you doubt his words, look to the first prophecy to the church as Peter (an apostle acting as a prophet).

Act 2:16 But this is that which was spoken by the prophet Joel:

Act 2:17 "And it shall be in the last days, says God, I will pour out of My Spirit upon all flesh. And your sons and your daughters shall prophesy, and your young men shall see visions, and your old men shall dream dreams.

106

Act 2:18 And in those days I will pour out My Spirit upon My slaves and My slave women, and they shall prophesy.

And permit me to digress, for those hard-hearted men who will tell their sisters that they have no place in ministry, read those scriptures at least three times, repent and turn from those wicked and carnal thoughts. I will use the masculine pronouns many times with the prophet, but just imagine the word prophetess and the female pronoun is there.

What is the role of the modern-day prophet – and they do exist? The same as it was in the days of Elijah, Amos, Joel, John the Baptist, and John, Peter and Paul to mention a few. Yes, I stepped on some toes of those who talk about the mythical and man-made five-fold ministry teaching which says a person can be a prophet, apostle, pastor, teacher, or evangelist. Let me dash that too. Jesus was all five and we are to be like him. We are to be what we are when we are needed to be it. Look at Paul's charge to Timothy. It shows teacher, evangelist, and prophet.

> 2Ti 4:2 preach the Word, be instant in season and out of season, reprove, rebuke, exhort with all long-suffering and doctrine.
>
> 2Ti 4:3 For a time will be when they will not endure sound doctrine, but they will heap up teachers to themselves according to their own lusts, tickling the ear.
>
> 2Ti 4:4 And they will turn away their ears from the truth and will be turned to myths.
>
> 2Ti 4:5 But you watch in all things, endure afflictions, do the work of an evangelist, fully carry out your ministry.

Few "pastors" I know who have been faithful to their calling have not been at least pastor, teacher, prophet and evangelist. I am not sure about the apostle, but I feel more than a few have been instrumental in setting doctrine, at least for their own church.

But let's look at those who God calls to be a prophet, to prophesy to his people. He will encounter attitudes of derision, disrespect, disdain and even sometimes physical threats. Allow me to show a few of them.

In I Kings 19 (cited above) Elijah had prophesied against the crown and now Jezebel has threatened that he will be killed. We could say to Elijah, "Man up." But before we do let us look at our own lives. I have with mine and there have been times that I have had a word from God for his people and it would have been a lot easier for me to be silent. To my discredit, there were times that I was. My experiences tell me that I am far from unique. I believe there are more than a few in the church who have heard from God, heard clearly, and to avoid the pain that they were sure would result have remained silent. We humans are good at working to avoid pain. The church hierarchy has silenced the prophet and in doing so, silenced God's word. It is no wonder the church is powerless today!

Amos is one who has both God's word on it and the personal experience of the attempt to silence. He alludes to it in this passage. In it he relates the word from God that He has raised up prophets and they polluted them and commanded them to not prophesy. In this case it is not the secular government but the people who have decreed it.

> Amo 2:11 And I raised up prophets from your sons and Nazarites from your young men. Is it not even so, O sons of Israel? says Jehovah.

> Amo 2:12 But you gave the Nazarites wine to drink, and commanded the prophets, saying, Do not prophesy.

Then Amos has proclaimed God's word which is the complaint against Israel. He shows they have polluted the ones sent and told others to not speak. The priest conspires against Amos with the king. The religious leader is now the accuser. Now the religious leader is the one who works to suppress the word of the Lord. In addition, he goes beyond the truth and accuses Amos of sedition. The priest has now joined Satan as the accuser of the brethren.

> Amo 7:10 Then Amaziah the priest of Bethel sent to Jeroboam king of Israel, saying, Amos has plotted against you in the midst of the house of Israel; the land is not able to bear all his words.

> Amo 7:11 For so Amos says: Jeroboam shall die by the sword, and Israel shall surely go into exile out of his land.

Amo 7:12 And Amaziah said to Amos, O seer, go, flee for yourself into the land of Judah; and eat bread there, and prophesy there.

Amo 7:13 But do not prophesy again any more at Bethel; for it is the king's temple, and it is the king's royal house.

We could look at verses 12 and 13 as the priest trying to protect the prophet by saying that he should avoid angering the king if he had not accused Amos of sedition.

Now we see here how Amos was called into the role of the prophet.

Amo 7:14 Then Amos answered and said to Amaziah: I was no prophet, nor was I a prophet's son. But I was a herdsman and a gatherer from sycamore trees.

Amo 7:15 And Jehovah took me from behind the flock, and Jehovah said to me, Go, prophesy to My people Israel.

And Amos's response is not what the priest wanted to hear and judgement is pronounced against him and his family. For attempting to silence the prophet the judgement is terrible.

Amo 7:16 Now then hear the Word of Jehovah. You say, Do not prophesy against Israel, and do not drop words against the house of Isaac.

Amo 7:17 So Jehovah says this: Your wife shall be a harlot in the city, and your sons and your daughters shall fall by the sword, and your land shall be divided by line. And you shall die in a defiled land; and Israel shall surely go into exile out of his land.

Amos and Elijah are not the only prophets that were silenced or ignored. One example here is Micaiah in dealing with Ahab.

2Ch 18:5 And the king of Israel gathered four hundred men of the prophets and said to them, Shall we go to Ramoth-gilead to battle, or shall I wait? And they said, Go up, for God will deliver it into the king's hand.

2Ch 18:6 But Jehoshaphat said, Is there not a prophet of Jehovah here besides, so that we might ask of Him?

Jehoshaphat here says something that puts the four hundred prophets in verse 5 into a perspective. He says, "Is there not a prophet of

Jehovah." He is certainly implying if not saying that the four hundred are not prophets of Jehovah. Let it be clear, if they are not prophets of Jehovah, they are false prophets. We have many prophets today that are not prophets of Jehovah, those who forecast weather, stock rise and fall, and elections. And if we look at their deplorable track records, we can see that the only thing we can validly place our trust is in God. His is the only voice that brings truth and only those who speak for Him are true. Sometimes the word from God for us it is that small voice that says, "This is not a good idea." Sometimes it is the voice of a prophet, and sometimes one who has not been certified by the official board of prophet certification. It may be one like Amos who would say, "I was no prophet, nor was I a prophet's son. But I was a herdsman and a gatherer from sycamore trees.' You can replace the herdsman and gatherer with modern day occupations. Jehoshaphat is wise enough to know where truth comes from and asks for that counsel.

2Ch 18:7 And the king of Israel said to Jehoshaphat, There is yet one man by whom we may inquire of Jehovah. But I hate him, for he never prophesied good to me, but always evil. The same is Micaiah the son of Imla. And Jehoshaphat said, Let not the king say so.

2Ch 18:8 And the king of Israel called to a certain eunuch and said, Bring quickly Micaiah the son of Imla.

2Ch 18:9 And the king of Israel and Jehoshaphat king of Judah sat each of them on his throne, clothed in robes. And they sat in a grain floor at the entrance of the gate of Samaria. And all the prophets prophesied before them.

2Ch 18:10 And Zedekiah the son of Chenaanah had made horns of iron for himself. And he said, So says Jehovah, With these you shall push Syria until they are crushed.

2Ch 18:11 And all the prophets prophesied so, saying, Go up to Ramoth-gilead and be blessed. For Jehovah shall deliver it into the king's hand.

2Ch 18:12 And the messenger who went to call Micaiah spoke to him, saying, Behold, the words of the prophets declare good to the king with one mouth, and please let your word be like one of theirs, and speak good.

2Ch 18:13 And Micaiah said, As the Lord lives, even what my God says, that I will speak.

We see the false prophets delivering their message and then the messenger tries to taint the word of the true prophet. I am sure he knew that the king would not be pleased with his message. But the true prophet answers that he will speak the word of the Lord. This is all a prophet should do.

2Ch 18:14 And he came to the king. And the king said to him, Micaiah, shall we go to Ramoth-gilead to battle, or shall I wait? And he said, Go up and be blessed. And they shall be delivered into your hand.

2Ch 18:15 And the king said to him, How many times shall I warn you that you say nothing but the truth to me in the name of Jehovah?

It is interesting here that the prophet tells the king what he wants to hear and the king realizes it is not the word of the Lord. The prophet then continues with the word of the Lord.

2Ch 18:16 And he said, I saw all Israel scattered on the mountains, like sheep that have no shepherd. And Jehovah said, These have no master; let them return, each man to his own house in peace.

2Ch 18:17 And the king of Israel said to Jehoshaphat, Did I not tell you he would not prophesy good concerning me, but evil?

The prophet now gives the true word of the Lord and the king responds implying the prophet is just saying it because he doesn't like the king.

2Ch 18:18 Again he said, And hear the Word of Jehovah. I saw Jehovah sitting on His throne, and all the host of Heaven were standing on His right hand and on His left.

2Ch 18:19 And Jehovah said, Who shall tempt Ahab king of Israel so that he may go up and fall at Ramoth-gilead? And one spoke saying in one way, and another saying in another way.

2Ch 18:20 And a spirit came out and stood before Jehovah and said, I will tempt him. And Jehovah said, With what?

2Ch 18:21 And he said, I will go out and be a lying spirit in the mouth of all his prophets. And Jehovah said, You shall tempt him, and you are able. Go out and do so.

2Ch 18:22 And now behold, Jehovah has put a lying spirit in the mouth of these your prophets, and Jehovah has spoken evil against you.

There is often an attempt to deride and deflect the word of the prophet. The prophet now gives the glimpse into what has happened in the heavenly realm. This is really the second warning against the action.

2Ch 18:23 And Zedekiah the son of Chenaanah came near and struck Micaiah on the cheek. And he said, Which way did the Spirit of Jehovah go from me to speak to you?

2Ch 18:24 And Micaiah said, Behold, you shall see on that day when you shall go into an inner room to hide yourself.

2Ch 18:25 And the king of Israel said, Take Micaiah, and carry him back to Amon the governor of the city, and to Joash the king's son.

2Ch 18:26 And you shall say, So says the king, Put this fellow in the prison, and feed him with bread of affliction and with water of affliction until I return in peace.

2Ch 18:27 And Micaiah said, If you certainly return in peace, then Jehovah has not spoken by me. And he said, Listen, all you people.

Ahab ignores the counsel and is killed in that battle.

Let us move on to another Old Covenant prophet, John the Baptist. Sure, he is in the New Testament, but the new covenant is not in force until the testator, Jesus is dead. Allow me to show that with scripture.

Heb 9:14 How much more shall the blood of Christ, who through the eternal Spirit offered himself without spot to God, purge your conscience from dead works to serve the living God?

112

> Heb 9:15 And for this cause he is the mediator of the new testament, that by means of death, for the redemption of the transgressions that were under the first testament, they which are called might receive the promise of eternal inheritance.

> Heb 9:16 For where a testament is, there must also of necessity be the death of the testator.

> Heb 9:17 For a testament is of force after men are dead: otherwise it is of no strength at all while the testator liveth.

The old covenant, Old Testament, the law was in effect until the death of Christ. That death ushered in the New Covenant.

John is probably one of the last Old Covenant prophets, of course Jesus is both the Son of God and probably the last Old Covenant prophet. We have no record I can find of one who prophecies after his time in the upper room.

As a postscript, statement "Let his blood be on us and our children" was prophetic. There may be a later one.

John the Baptist is not a part of the religious hierarchy. He is an outsider. He offends the leaders and the civil authorities by pointing out their sin. He calls for repentance. The masses are moved to "make way for the Lord" but the religious leaders are satisfied with the status quo that holds them in places of honor, not their rightful place of service to God and the people.

I will share something here, like Amos, I am not a prophet nor the son of a prophet. I am just a writer computer programs and network scripts. But I believe there is a message from God to the church in what I am writing, a message that is necessary and I am by far not the only one he has given this to nor am I the only one presenting it. It is the same message that John the Baptist brought.

All too often the message has come from outsiders in the modern church and has been in the church for as far back as my memory and research takes me. Where it has been from inside it has been from the bottom up, not the top down in the hierarchy. Allow me to pose this. Let's suppose a school janitor, truck driver, computer programmer, print shop worker, or the like who is without reproach in the community and the church stood in a service and said, "I have a word from the Lord for this body." Would he, a. Be told to sit down, b. Escorted out, c. Booed, d. Denounced from the pulpit, e. Asked to share it privately with the pastor first, or f. Told to share it?

As an aside I see nothing wrong with e. if it is used to be sure the prophecy is for the church and is prudent unless the person has been doing this and has been responsible with handling the word. However, the pastor or if he chooses a panel of elders must simply see if it is what he proports and then let him speak and in a timely fashion. God's word for now is for now, not next week or after the board meets. Unfortunately, I would bet if this happened 100 times, 99 of those would not result in either e. or f. except in a few churches where this has been supported by the leadership.

Allow me to look back to John the Baptist and Amos. The scenario on the church is very near to the display of their prophetic voice. The only question is whether this is appropriate under the New Covenant. Although the word covenant and testament are somewhat the same, I use the two words to separate the two divisions of the bible (Testament) and the two covenants from God.

Christ had to die to usher in the new covenant therefore all before his death was under the old. This one thing is essential for us to understand and teach the things of God. We must understand and rightly divide the word of truth.

> 2Ti 2:15 Study to shew thyself approved unto God, a workman that needeth not to be ashamed, rightly dividing the word of truth.

Then what is rightly dividing the word of truth? As I stated, the most important division is the boundary of the cross. It is the point on the time line of human history where man's relationship with God changed. There was one other such pivotal moment, when Adam sinned. In Geometry we call an intersecting line at 90 degrees a perpendicular bisector, a line that divides another line into two parts and is at 90 degrees to the line. To me, that aptly describes the cross on that time line. The Greek word 'orthotomeō' which only appears in this verse derives from the Greek word 'orthos'.

orthos

or-thos'

Probably from the base of G3735; right (as rising), that is, (perpendicularly) erect (figuratively honest), or (horizontally) level or direct: - straight, upright.

Bill and Gloria Gather expressed it a little different but just as validly. (Selected verses and chorus)

> There's a line that is drawn through the ages
> On that line stands an old rugged cross
> On that cross, a battle is raging
> To gain a man's soul or it's loss
>
> On one side, march the forces of evil
> All the demons, all the devils of hell
> On the other, the angels of glory
> And they meet on Golgotha's hill
>
> The earth shakes with the force of the conflict
> And the sun refuses to shine
> For there hangs God's son, in the balance
> And then through the darkness he cries
>
> It is finished, the battle is over
> It is finished, there'll be no more war
> It is finished, the end of the conflict
> It is finished and Jesus is Lord

Yes, the work of the cross was finished. On one side is the law. On the other side of the line is Salvation, Grace and Forgiveness through the blood of Christ. We have the right to accept or reject.

But let's go back to the New Covenant prophets. There are few references to prophets, but they are there. One of the most poignant is in the Acts 2, the beginning of the church. Peter reaches back across the line to the old covenant, to the prophecy and talks about it being fulfilled.

> Act 2:16 But this is that which was spoken by the prophet Joel:
>
> Act 2:17 "And it shall be in the last days, says God, I will pour out of My Spirit upon all flesh. And your sons and your

daughters shall prophesy, and your young men shall see visions, and your old men shall dream dreams.

Act 2:18 And in those days I will pour out My Spirit upon My slaves and My slave women, and they shall prophesy.

Act 2:19 And I will give wonders in the heaven above, and miracles on the earth below, blood and fire and vapor of smoke.

Act 2:20 The sun shall be turned into darkness and the moon into blood, before that great and glorious Day of the Lord.

Note several things. Firstly. Your sons and daughters will prophecy. They will proclaim the Word of the Lord. Not only the sons, but also the daughters. On this foundational prophecy of the church I will oppose those who say our daughters cannot be full participants in the work of the ministry. I need nothing more but there is more support elsewhere. Secondly. On this foundational statement we see that they will prophesy. Will that end at some point? No. It will continue till the great and glorious day of the Lord. It will continue till the church comes into the unity of the faith.

Eph 4:11 And he gave some, apostles; and some, prophets; and some, evangelists; and some, pastors and teachers;

Eph 4:12 For the perfecting of the saints, for the work of the ministry, for the edifying of the body of Christ:

Eph 4:13 Till we all come in the unity of the faith, and of the knowledge of the Son of God, unto a perfect man, unto the measure of the stature of the fulness of Christ:

God's word and power will be proclaimed. Thirdly, it does not say the priests, the pastors, the evangelists will prophesy. It says, your sons and daughters. That can include the clergy but is not limited to it. If history holds the answer to that, the clergy has rarely been the vehicle for prophecy to the church that defines a change in direction. Look at young Samuel. He heard the word of the Lord and proclaimed it to Eli. As a young boy he was a prophet! Too many of the church leaders' statements are political statements, not prophecy. Too many have actually opposed the presentation of the word of the Lord.

116

Let's look at some of the examples in the New Covenant. The one that comes to mind first is John in the Revelation of Jesus Christ. Here he both presents the word of the Lord, he also talks of things to come. It is the Revelation of Jesus Christ, not revelations as many call it – something that may seem minor but it hides the fact that Christ is revealed in this book.

> Rev 1:1 A Revelation of Jesus Christ, which God gave to Him to declare to His servants things which must shortly come to pass. And He signified it by sending His angel to His servant John,

> Rev 1:2 who bore record of the Word of God and of the testimony of Jesus Christ and of all the things that he saw.

The words John wrote tell of the future and tell us about Christ. But even more they are warnings to the church today. I have heard arguments till I was sick about whether these churches were states of the church, churches existing in John's day or to the church today. I don't care. If we spent the time working to bring people to Christ that we have wasted on such trivia, the church would be a force in the world today. I know this.

> 2Ti 3:16 All Scripture is God-breathed, and is profitable for doctrine, for reproof, for correction, for instruction in righteousness,

> 2Ti 3:17 that the man of God may be perfected, thoroughly furnished to every good work.

I know those portions are for me today to be more vigilant and prepared. I also know that I will not know the day or hour based on the gospels and I am to be going about His work not speculating on when he is coming. Not one scripture tells me to do that. There are many that tell me to avoid vain words.

Apart from John most would dispute the prophecy under the new covenant. There are so many times that the word of the Lord goes forth but there are more than a few places where the word includes the future.

> Act 21:10 And as we stayed more days, a certain prophet from Judea named Agabus came down.

Act 21:11 And coming to us, and taking Paul's belt, and binding his hands and feet, he said, The Holy Spirit says these things: So shall the Jews at Jerusalem bind the man whose belt this is, and will deliver him into the hands of the nations.

Act 21:12 And when we heard these things, both we and those of the place begged him not to go up to Jerusalem.

Act 21:13 Then Paul answered, What are you doing, weeping and breaking my heart? For I am ready not only to be bound, but also to die at Jerusalem for the name of the Lord Jesus.

Act 21:14 And he not being persuaded, we ceased, saying, The will of the Lord be done.

Act 21:15 And after those days, making ready, we went up to Jerusalem.

Note that the word was to prepare Paul for what was to come, not to deter him from doing God's will out of fear.

Paul talks about prophesying more than a few times, both himself and others. I could go into more in both the old and new covenant but let us come to today.

Are there prophets today?

I will answer that in the affirmative. I know because I have seen and heard them. There are. I am also going to address the counterfeits that are many and show some of how to identify them. Firstly, when God gives a word through a prophet, it is clear, it is to the point and those for whom it is directed KNOW. They may operate in open denial but down deep they know. Secondly, there is no need for someone to interpret or explain the prophecy that is given. Tongues need to be interpreted, prophecies do not. If someone starts to interpret prophecies, pay no mind to them. They are polluting God's word with the thoughts of man! Polluting is a harsh word, but it is the correct one. Thirdly there are men who travel and profess to give the word of the Lord to people, both individually and to groups. In this I have seen the good, the bad, and the ugly. Regrettably I have seen the bad and the ugly far too often. I refer to many of these who fall into the latter category as Christian Fortune Tellers. When I see these, I look at the fruit, what comes from their ministry, be it good or evil. As an aside, I have seen the same prophet in one setting give prophecies that were 100% right on and some others that were to scratch itching ears. One such prophet on one night gave a word over a man who had done some terrible things and it was one that seemed to many in the church that it was a word of encouragement. I knew the situation and Dee and I were counselors at the time. A distraught mother of the girl who was harmed by the man exited the sanctuary in tears. Dee and I exited behind her. She was sobbing, questioning how this word could be given to him. I asked her to recite the word as she remembered it. When she did, I showed her it was not encouragement but a statement of where he could have been had he not done the wrong acts. I was in the raised sound are and could see the man's face when the prophecy was given. I assured her, he knew exactly what the Holy Spirit was revealing. The Holy Spirit was able to rebuke the man so softly that very few, possibly only him, Dee and I saw the real meaning. I believe we only saw it to be able to comfort the hurting mom.

In another service the same man called me up to 'give me a word'. He started, "I see great disappointment in you." I at first felt this was a rebuke to me from God. He paused for some time and I realized it was not God saying, "I am disappointed in you", but "I see great disappointment in you." My heart was at the time sorely disappointed, but the source of that disappointment was the man standing in front of me, the one giving me that word. I don't remember the remainder of that word other than that it was from God and it was confirming His love for me. I had a personal encounter with the man earlier in the day and saw him verbally abuse his wife in a way that I was sure was comfortable for him to do   As I spent time that day, I saw earmarks of abuse in the wife. One thing for sure, God showed me he could work through flawed vessels. I am very sure that during the pauses God was speaking to him.

We had a couple who always showed up every time that man was here. They were affluent, contributed to his ministry and always were called up to get a 'word from the Lord.' I will let you sort that one out. I will only say that I know there were prophecies he gave that were right on, were from the Lord.

There is another group that will take a something they have written or spoken in the past and show how it fits an event, usually an election or a disaster. I have heard more than a few of these and they, like the predictions of Nostradamus are so vague and meaningless that although I am careful to not disparage the word of the Lord, I can discount them when they are not clear. Most of the time I have seen an uncertain prophecy it has been from one who is not mature or one who has tried to soften the word. Please, if God gives you something, pass it on.

A few months after Katrina there was an e-mail going around that one of these men had predicted the storm several months before it hit. I found the recording of the prophecy. I listened to it several times. Not in my wildest imaginations would I have been able to define it as the prediction of a devastating storm and there was no pointing to the location.

I will simply say that if the word is not clear enough to be correction, encouragement, warning, or a call to action, and does not lead to a better understanding of God it is not of God. In addition to the scripture above that says all scripture is for our benefit, God defines how he instructed Habakkuk to write his word.

> Hab 2:2 And Jehovah answered me and said, Write the vision, and make it plain on the tablets, that he who reads it may run.

The words of the Prophets of the Old Covenant were clear, to the point and direct. John the Baptist's words were the same to the extent that they caused his death. Stephen, certainly one of the New Covenant prophets was stoned for presenting God's word. And although most modern prophets are not stoned, many are shunned, and attempts are made to silence them by any means possible. Been there, done that. I know first-hand. Dee and I were feared enough that one pastor, from the pulpit, told the congregation it would be best if they do not have contact with us.

What of the false prophet, then and now? I will show the word of the Lord on this.

> Eze 13:1 And the Word of Jehovah came to me, saying,

Eze 13:2 Son of man, prophesy against the prophets of Israel who prophesy. And say to those who prophesy out of their own hearts, Hear the Word of Jehovah:

Eze 13:3 So says the Lord Jehovah, Woe to the foolish prophets who follow their own spirit and have seen nothing!

Eze 13:4 O Israel, your prophets are like the foxes in the deserts.

Eze 13:5 You have not gone up into the breaks, nor built the wall for the house of Israel, that it might stand in the battle in the day of Jehovah.

Eze 13:6 They have seen vanity and lying divination, saying, Jehovah says. And Jehovah has not sent them; but they hoped to confirm their word.

Eze 13:7 Did you not see a vain vision, and speak a lying divination? Yet you say, Jehovah says; although I have not spoken?

Eze 13:8 Therefore so says the Lord Jehovah: Because you have spoken vanity and seen a lie, therefore, behold, I am against you, says the Lord Jehovah.

Eze 13:9 And My hand shall be against the prophets who see vanity and who divine a lie. They shall not be in the council of My people, nor shall they be written in the writing of the house of Israel, nor shall they enter into the land of Israel. And you shall know that I am the Lord Jehovah.

Eze 13:10 Because, even because they made My people go astray, saying, Peace; and there was no peace; and he builds a wall, and lo, others daubed it with lime.

There are prophets today. Unfortunately, some, in fact too many are false prophets. We are to be as wise as serpents and as harmless as doves. It is incumbent on those who speak for God to accurately express His will and ways, not taint them with their own ideas or the norms of the society. I will admit that this is a difficult and continual task and no matter how diligent, we will fail at times. It is incumbent on those who hear the prophecies to have the wisdom from God to know who and what is of Him.

I fear that many of those who are professing to be prophets today are not Isaiah's but there are many Amos and Habakkuk class prophets today who are in their own venue, presenting the word of God accurately.

Allow me to share from experience. I have been one who several times God has given a word in and for a specific time and place. With exception of those that will serve as examples I will not share them here because they are not for this time or place.

I will share several here, one because it shows how we can fail God and our fellow man. One evening in a service we had a well-known speaker. As he was speaking, I started hearing and writing what I was sure was from God. I write if possible, what I hear. I don't want to taint it.

I missed being able to hand it to him before he left for the evening but was aware that the pastor was going to take him to the airport Monday morning. I handed it to the pastor and asked that he give it to the man. He apparently felt I had no right to speak into that man's life, didn't see an issue and he didn't pass it on and he did not tell me he wasn't going to. I may have tried to get it to the man had I knew that, but I can't be sure. I may have wimped out. The prophecy simply said that his marriage was very brittle, could easily be in trouble and he needed to concentrate on the relationship and spend more time with his wife to avoid problems.

About two months later our pastor and his wife were called to help the couple work out marriage problems. The wife had an affair. I had the word, I failed to deliver it. I could have found which hotel he was in and delivered it. I didn't give the man God's best. Neither did the pastor. The pastor never mentioned the note after the problem. A 'we screwed up' statement would have raised my respect for him and removed a cloud.

They are different from the teachings I write because they are specific and call for a specific action. What God gives me as teachings are far more general and may have value beyond that time and place. Every time I have had such a word it has been in a place where correction or encouragement was sorely needed. God has told me he has called me as a watchman in his church. Let me tell you, a watchman is not a beloved creature. He is the one who blows the trumpet at 3 AM when the army is sleeping and enemy is advancing.

It is the call to rise and fight. The watchman will never be met with joy and gladness.

When my wife and I were installed as elders in a church one of the men spoke prophetically over us. For over a month before I was dealing with the word watchman. God kept bringing it to mind and I had read every scripture on it several times. I knew what it meant and he stated that I would be a watchman in the House of God.

I had no doubt of the accuracy of that word from the Lord. I listen to exact wordings in prophecies. Later I realized it was in the House of God, not just that church. Had it been a specific church I would have considered that mandate to end when we left that church. It didn't.

After that service, I told him that was correct but not a good word for me. He asked why I would say that. I responded, "The watchman is the one nobody likes because he disturbs their sleep."

That assessment was borne out in less than 4 years when in concert with the other elders we questioned something and we were all fired. When the word is there and it is the time to give it, there is only one way to avoid it. Paul said it when he was before Agrippa when he gave the word of the Lord to him.

> Act 26:19 Whereupon, O king Agrippa, I was not disobedient unto the heavenly vision:

The one like Amos and Habakkuk, who is entrusted by the Lord to give His word has a choice. As Yogi Berra said, "when you come to a fork in the road, take it." The correct fork is the one that is being obedient to God, something that should be self-evident but there are pressures to take the other. He can give the word or be disobedient. I have often heard preachers talk about preaching without regard for fear or favor of men. I have laughed as I heard it. Many of them have no idea what that is to face men who do not want to hear what you are speaking. Most pastors work to a friendly audience. Most of the time the prophet is delivering God's word to a hostile or at least unhappy one.

But I will go back to what I said, the one who God calls as a prophet at that time and place can either give his word or be disobedient. I must admit I have backed up, a nice way of saying, being disobedient at times. I have most times given His word. And I at times have suffered for being obedient. But as Peter said before the religious court, "We ought to obey God rather than man." I will tell you the suffering for obedience is far easier to handle than the knowledge that you failed to do what God called you to do. There is a song, "So many lives depend on what I do." As I pass my three quarters of a century here on earth, I look back and see more how my life has impacted others, some good, some bad. I am as sure I co not see all of them.

I have related the story of my dad giving up a chair for a young mom with a baby, the couple staying in the service and giving their hearts to the Lord that night, them going on for God, him becoming a pastor and impacting directly, literally several thousand lives. I have often asked, "what if?" I am sure there are thousands of other stories like that out there. They can either be an act, a word, whatever, but they impact lives.

DIVISION IN THE CHURCH

I have been in the church for over seventy years. I have seen church divisions from near and afar, in the church I attend and in other churches. They are never pretty sights. I have helped pull things back together after them. At times I have been sidelined by the leadership when things were going well, till the problem hit, then I am desired and called in to help clean up the mess. I somewhat feel like the school janitor who is called when messes occur.

I will be frank. I never want to see one of these again. But most of the divisions I studied up till a few years ago were vertical splits where a portion of the congregation split off from the rest. The fracture was vertical. And I never looked at any other possibility.

To set the stage, let's look at the anatomy of a church split. Luther's split with the Catholic Church was vertical. The split between the Greek and Eastern Orthodox was vertical. Organizations are pyramids with a leader on the top and the rank and file on the bottom. Think of a vertical split as a portion of the pyramid split off and moved away. The size of the fragment can be small or large. The tip of the new pyramid can be low or high but the higher it is, the larger the segment that moves away. With a vertical split there is some leader that takes the people away. This can be a formal or informal leader. I have several times seen it be an associate pastor or worship leader.

A good pastor recognizes the existence of the informal leaders in the church, those without titles who because of their character and leadership ability are looked up to by others. He utilizes them just as he does formal leaders. If he does not, he can have serious problems. A split can be led by formal or informal leadership. Rarely does the informal leader see his actions as potentially resulting in a split.

But what happens when the spilt is horizontal? Up till now I have not looked at it in this light and hadn't thought of it. I have actually seen at least a half dozen of these over the years and only now am seeing what happens. I started seeing this while in church several weeks ago. It was prompted by something someone did that I mentally said, "This could cause a split." Then I realized it would not fit the paradigm of the usual split I understood. I started looking at how this would impact.

One of the symptoms of a horizontal split is a physical loss of membership or at least loss of a significant number being a vital part of the church. The latter is often as devastating as the actual physical loss. Sometimes the loss is greater than the numbers would indicate/ One of the first signs of this is seeing committed members slip to pew sitter status or leave. Sometimes they just leave and sometimes they move through two steps.

The typical answers to this by leadership are, "They were never really with us." Or "They were never happy here." It is easy for leaders to take this approach because these people are often the ones who commit to something but want to be sure it is going right so they are problems because they ask too many questions. It is almost easy to look at their leaving as a good thing until too many of them leave. These people rarely make a big fuss when they leave, they just disappear. Sometimes the only way anyone notices is that holes appear in the ranks of workers. It is rare that more than a few move to the same church. They appear randomly in other churches. They never form a new church. The less committed are less impacted by the causal agents of the split so they rarely leave.

What happened? What does it look like? Can I recognize it? Look at this scenario. Over three months you have lost up to a fifth of your most committed people, an insignificant number compared to the whole congregation and nobody is really saying anything is wrong. In a congregation of two hundred the real committed ones are no more than thirty, one fifth of them is six. There may be a few people in the next level of commitment that are either skipping services or have gone too. It is harder to tell with these because they sometimes miss a couple of services in a row. You write it off to bad weather or vacations but forget these people come to church in nearly every kind of weather and they don't take that many Sundays on vacation.

If you are a pastor and have seen this happen, you are in trouble but be of good cheer. The problem can be fixed. Don't fret the ones who have left. It is like blood that has come out of a wound. The losses you have sustained are permanent, you can't put the blood back in, but the hemorrhage can be stemmed with the right action. What is critical here is the same as with a wounded person, stop the bleeding. And what you are seeing is a church that is bleeding. Let's look at how this happens.

If this has happened in your church, you have just experienced a horizontal split in the church. And I will be willing to bet a twenty that nobody in seminary every told you this existed, what causes it or how to handle it if it does.

How is it that a significant group can decide to leave in a short time without a leader to take them away? There has to be a trigger. King Saul had one committed soldier on the field of battle. He could have lost that soldier had he not connected with David. The lad asked a question. "Is there not a cause?" Saul answered that and there was an instant connection between the leader and the one led. David who had seen the need to fight was now seeing that he was being given the nod to do it. The most committed people in your church are looking for two things from you, the word that there is a cause and the nod to go. There was no horizontal split on Flight 93 when Beamer said, "Let's roll." They knew they were in it together. There was no elite, no royalty. Beamer was part of the group. The leader was not issuing orders. He was in the thick of it with them. They accomplished their mission, Flight 93 ended with only those on board being killed compared to the 3000 dead at New York. I am sure they hoped they would survive but most important, this had to stop.

A horizontal split occurs when the leaders move away from and loose connection with the rest of the church. The split is between the leaders and the church and it starts out as a lack of connection. Some will tolerate this but the most committed are the ones most impacted. Jesus said his desire was that they would be one. This not only means that the congregation is one or the leadership is one, it is that together they are one. Why is it so amazing to us that when we do it God's way it works and when we leave that way it doesn't?

What causes a horizontal split or better, what can we do to avoid one and avoiding is the best plan? <u>We can never allow the leadership and the congregation to lose contact.</u>

Give me a couple wedges, a heavy hammer and an axe I can split an oak log. I have done it. The Axe will not split it. But I can get a little notch with the axe, just a little one in the end of the log. I set the one wedge in that notch and I drive it with the hammer. When it has hit its maximum gap, I set the other wedge into a narrow place in the crack I have created. I drive that wedge further opening the crack. The first wedge falls free and I repeat the process. I can rend a solid log into pieces.

A horizontal split in the church has the same anatomy. You get a little notch. You drive a wedge. You set another wedge and widen the split. The notch is usually subtle. It is something that pulls the congregation away from the leaders. Often this notch is seen for what it is by the leaders, a mistake but rather than admit it they try to justify it and the wedge is driven. The congregation looks at it as, "They don't trust us enough to tell us." A second notch is some form of separation. Leaders in the church are there to serve the people. The big brother and sister view is great. They are still a part of the body. What can get divisive is the 'set apart' idea. The moment you do this you have cut the notch. Now all Satan has to do is apply the wedge and hammer. Little things become big. It is the little foxes that spoil the vine.

I cringe when I see anything that sets leadership apart from the church and I did when I was in leadership. The only honor I sought was the confidence the pastor and the people had in me to be able to serve them. Anything more than that is going to be destructive.

And something so seemingly good as pressing for prayer for the leadership rather than for the whole church can be destructive. It sets them apart. It can be the notch. I have actually seen this play out in one situation! Dee and I only asked for special prayer for us in a leadership role in one situation. We were prayer ministers and when we started counselling with a new person or couple, we asked two people who prayed for us to be in prayer for us that we would minister effectively and for that one who was being counselled. On the other hand, in one church a special prayer group was created to pray for the pastor. It created a crack.

Beware of honor for leaders. Beware of anything that sets them apart from the people.

Copyright Ralph Brandt 2009

ABOUT THE AUTHOR

I believe it is well within reason to ask, "Who is this? What right has he to write?"

I have been authorized like every Christian, everyone who has committed his or her life to Christ, to present His word. Paul says to be instant in season and out of season. This is something that is beyond the scope of men to override. I challenge you to decide, to discern, to seek God and ask, "Is the Holy Spirit present in what I have written?" I challenge you to judge this work.

I tell people I was born in the fire and I can't live in the smoke. I was in Pentecostal tent meetings in the 1940's and 1950's. I have seen God's power, His healing, His saving power in action. I have seen lives changed. I have been active in churches except for only a very few months in my life. I have been shunned out of a church because I held to the right. About 10 years ago after the second bad event, I vowed that I would never, never again be in leadership in a church. As far as I was concerned, that was it. But God had another idea. But more important to me than the church is the relationship with Him. It is the communion of the Holy Spirit. But let us not forget or neglect also the fellowship of the saints, something all too many reject. Believe me, I understand why some, particularly those who want to follow God explicitly will find times that they will want to walk away from what is called the church. But imperfect as it is, the scripture says that where two or three are gathered, He is there. Yes, my relationship with Him is personal, but it is also lodged in a corporate setting. I can worship him anywhere. I know I could worship him, even in Sardis.

www.ingramcontent.com/pod-product-compliance
Lightning Source LLC
Chambersburg PA
CBHW021828170526
45157CB00007B/2718